WEIRD CANADIAN ANIMALS

Fascinating, Bizarre and Astonishing Facts from Canada's Animal Kingdom

WEIRD CANADIAN ANIMALS

Fascinating, Bizarre and Astonishing
Facts from Canada's Animal Kingdom

Wendy Pirk

**BLUE
BIKE
BOOKS**

The Publisher: Blue Bike Books
Website: www.bluebikebooks.com

Library and Archives Canada Cataloguing in Publication

Pirk, Wendy, 1973–
  Weird Canadian animals : fascinating, bizarre and astonishing facts from Canada's animal kingdom / by Wendy Pirk.

Includes bibliographical references.

ISBN 978-1-897278-52-9

  1. Animals–Canada–Miscellanea. I. Title.

QL50.P57 2009   591.971   C2009-900187-X

Project Director: Nicholle Carrière
Project Editor: Nicholle Carrière
Cover Image: Photos.com
Illustrations: Peter Tyler and Roger Garcia
Photography Credits: Every effort has been made to accurately credit the sources of photographs. Any errors or omissions should be reported directly to the publisher for correction in future editions. Photographs courtesy of © Alain | Dreamstime.com (p. 44); © Bridgetjones | Dreamstime.com (p. 39); © Buch | Dreamstime.com (p. 73); © Cascoly | Dreamstime.com (p. 157); © Dblais4 | Dreamstime.com (p. 191); Terry Goss / Wikipedia.org (p. 212); © Pali | Dreamstime.com (p. 141); pfly / Wikipedia.org (p. 27); Photos.com / Jupiterimages (pp. 11, 29, 59, 61, 78, 81, 94, 96, 105, 107, 116, 122, 178); © Sekernas | Dreamstime.com (p. 167); dominic sherony / Wikipedia.org (p. 65); U.S. NPS / Wikipedia.org (p. 166); unknown / Wikipedia.org (p. 195); Elaine R. Wilson / Wikipedia.org (p. 16).

We acknowledge the support of the Alberta Foundation for the Arts for our publishing program.

We acknowledge the financial support of the Government of Canada through the Book Publishing Industry Development Program (BPIDP) for our publishing activities.

Canadian   Patrimoine
Heritage   canadien

PC: 1

DEDICATION

To my family, as always, for their love and support. And to Elaine. If the world were full of people like you, it would be a far better place.

CONTENTS

INTRODUCTION. 9

WEIRD BY DESIGN
A Face Only a Mother Could Love 11
Structurally Strange 21
Flawed Design 31

FUNKY PHYSICAL ADAPTATIONS
Form Follows Function 33

SURVIVAL STRATEGIES
Chemical 48
Other Body Adaptations 55
Camouflage and Mimicry 59
Behaviour. 65

HUNTING STRATEGIES AND ADAPTATIONS
The Better to Catch You With 70

WEIRD FOODS
You're Gonna Eat That? 80

STRANGE BEHAVIOUR
Tool Use. 97
Wonderfully Weird Behaviour 100
Grooming and Self-medicating 102
Not Getting Along. 105

COMMUNICATION
Socializing, Animal Style. .107

MEETING, BREEDING AND REARING
Looking for Love 118
Reproduction 126
Bringing Up Baby 132

GETTING AROUND
I Like to Move It, Move It.....................138
Migration................................146

HOME, STRANGE HOME
Unconventional Dwellings......................151

OUT OF ITS ELEMENT
Like a Fish Out of Water.....................158
Urban Animals 162

UNLIKELY ASSOCIATIONS
Mutually Beneficial 165
A Bit More One-Sided.......................169
Unexpected Associations 173

A MISH-MASH OF WEIRDNESS
Further Factoids............................175

ALIENS AMONG US
Under Siege!................................185
Uninvited Imports 187
Unexpected Guests..........................193

CANADIAN CRYPTIDS
Is Anyone Out There? 196
Strange Hybrids............................201

ALREADY GONE
Nice Knowin' Ya 205
At the Hand of Humans 209
On the Brink...............................211

ACKNOWLEDGEMENTS

Thanks to Nicholle for her patience and input.

INTRODUCTION

There was a time when people believed that the world was full of magical, fantastic creatures. Dragons, griffons, horrifying sea serpents, unicorns and the phoenix were as real to the people of the past as bears, vultures and whales are to us today. Compendiums of nature, called bestiaries, that were created in medieval times still exist and recount the behaviour and natural history of creatures as the great thinkers of the day understood them. Thanks to the discoveries of science, much of the information contained in these books has been disproved, and some even seems so silly to our modern mentality that it is difficult to believe people could have ever accepted the information as true. However, some of the information in these texts that was discounted by later naturalists as too weird to be believable has actually come back into favour as science continues to provide us with a better understanding of the animals that share our world. And some of these creatures are weird, indeed. I'm not talking about exotic creatures from faraway jungles, either. I'm talking about our good, old Canadian animal species—many of which you might see in your backyard or during a stroll in the woods or along the shoreline.

So what makes these creatures weird? Well, to answer that, we must first determine exactly what "weird" means. A dictionary definition gives the meaning as "something bizarre or very strange," but that doesn't really tell us much. The more something strays from the boundaries of our understanding or our common experience, the stranger or more bizarre it seems, so perhaps a more apt description would be "something that is not like us." With this description, then, a weird animal must be one that looks or behaves in ways that are far removed from our own appearance or behaviour, right? Well, partly, but this is only part of the story. It seems that with animals, weird can also sometimes mean just the opposite. We humans like to think of ourselves as unique in the animal kingdom, and when we see something non-human displaying behaviour that we like to think of as especially "human," it may also strike us as weird.

So basically, it seems that where animals are concerned, weird can mean anything from unexpected or unique to truly bizarre or fascinating and, as you will see when you read on, many Canadian animals definitely fit this description.

A FACE ONLY A MOTHER COULD LOVE

Phantoms of the Abyss

We know they are out there. We even know where to look for them—well, sort of.

As far as we can tell, giant squid are widespread; they can be found in all of the world's oceans, usually near continental and island slopes. But knowing this hasn't made the creatures any easier to find. The giant squid has been the stuff of legend since the first ones washed ashore on the beaches of Scandinavia in medieval times, but even in ancient times, people spoke of giant, tentacled monsters that could attack and cripple boats at sea. Were they describing the giant squid? Some scientists think so. But as with so many things about this creature, we just don't know. So what do we know? Not much. The only concrete evidence we have of the giant squid comes from a few washed up carcasses—about 300 specimens (ranging from a few detached tentacles to entire bodies)—that have been hauled up as bycatch in fishing nets and a few bits and pieces fished out of the stomachs of dead sperm whales. There were no photos of the squid in its natural environment until 2005, when Japanese researchers caught one of the giant beasts on film as it attacked some baited lures.

Scientists believe the giant squid can grow to almost 18 metres in length. It has circular, toothed suction cups on its tentacles, and its eyes can be as large as soccer balls. Researchers had assumed that the squid hung out at depths of about 1000 metres, but new evidence suggests that, though the creature most likely travels throughout the water column, it spends most of its time in much shallower water than previously believed, perhaps no more than about 300 metres. A number of giant squid species have been discovered, including *Architheuthis sanctipauli* (a southern hemisphere species), *A. japonica* (from the North Pacific) and *A. dux* (found in the North Atlantic). Most sightings of *A. dux* have occurred in the North Atlantic Ocean, especially off the coast of Newfoundland, Norway and the northern British Isles.

DID YOU  KNOW?

In February 2007, a group of New Zealand fishermen caught a massive squid off the coast of Antarctica. The creature,

Mesonychoteuthis hamiltoni, otherwise known as the colossal squid, measured almost 10 metres in length and weighed more than 495 kilograms. It is one of only six colossal squid specimens that have ever been found, and it is the only one that is fully intact. The first evidence of this species' existence came in the form of two massive arms that were found in a sperm whale's stomach in 1925. The most recent catch has been preserved and is now on display at Te Papa Tongarewa, the national museum of New Zealand.

Elephants of the Sea

Imagine that you were fashioning a massive seal out of play-dough, but when it came to shaping the creature's face, you pulled a little too hard on the dough around the nose, and it

stretched so that it flopped down over the seal's face. That, essentially, is what the male northern elephant seal looks like. His droopy nose, or proboscis, begins to develop when he reaches sexual maturity at about three years of age, and by the time he is nine, the nose has reached its full floppy glory. The proboscis has been likened to an elephant's trunk, which is why this creature is named the elephant seal.

Bubble-nose

The hooded seal is so named because of the male's enlarged nasal cavity, called a hood (obviously). The hood, which begins to develop when the male is around four years old, has two lobes that the seal can fill with air. When inflated, the hood can be larger than a football, giving the seal a bulgy-looking forehead; when not puffed up with air, it hangs down to the seal's upper lip.

However, as weird as it is, the hood is not this seal's strangest feature. That honour goes to the inflatable, red membrane that the male seal has in its nose. The male inflates this membrane during courtship displays or to show dominance, and when inflated, the "balloon" protrudes from one of his nostrils. Basically, it looks as though the seal was chewing a piece of bright red bubble gum and blew a giant bubble out his nose.

The hooded seal is an Arctic species and can also be found in the northern Atlantic Ocean.

Unicorn of the Deep

If you should ever be cruising through Arctic waters or along the northern Atlantic Coast, keep your eyes peeled for a male narwhal. It'll be the one with the long, spiralled tusk jutting out of its face. Rather hard to miss, really. The tusk is actually an overgrown incisor that grows from the left side of the male's upper lip; the female occasionally grows a tusk of her own, but it is usually small, thin and weak—a poor specimen when compared to the spear-like tusk that grows on the male. Occasionally, a male will grow a second tusk from the right side of its mouth, but this, too, is usually a sorry effort, pretty much on par with the female's puny tusk. Scientists suggest that the tusk is spiralled to ensure that it grows straight as it protrudes from the whale's mouth. The theory is that a straight tusk will not burden the whale as it swims. Now, I'm no expert, but it seems to me that a 3-metre-long tooth growing out from the face and weighing as much as 10 kilograms would still be a bit unwieldy, now matter how straight it was. No one is sure what the whale uses this massive tooth for, but some proposed theories suggest that it is used to spear food (this seems incredibly unlikely—how would the whale then get the food to its mouth? His short little flippers would certainly be of no help), for self-defence or to chisel breathing holes into thick ice. The theory currently in vogue is that the male uses his tusk to duel with other males over mating rights and to establish dominance hierarchies, which, I guess, makes the tusk the cetacean equivalent of a ram's horn.

> **cetacean:** a marine mammal from the Order Cetacea, with a dorsal blowhole, fin-like forelimbs and no hind limbs; includes dolphins, whales and porpoises

In medieval Europe, the narwhal's tusk was passed off as the horn of a unicorn. Because it was believed to have come from a magical creature, the tusk was thought to have healing properties, and it was much sought after.

You've Got a Little Something on Your Nose

You know what it's like when you're taking a drink, and you either sneeze or someone makes you laugh, so the liquid comes shooting out of your nose? Well, that sort of describes the face of the star-nosed mole, only instead of "liquid," think "worms," and you should get a pretty good visual of what the mole's nose looks like. This peculiar-looking mole has 22 fleshy projections at the end of its snout. They are positioned in such a way that they form a ring around each nostril. The two middle projections located at the top of the snout remain still and point forward, whereas the rest of the fleshy nubs are in constant motion, feeling around for something to eat. These rather creepy-looking projections are highly sensitive and are much more effective than the mole's weak eyesight for locating potential prey, both on land and in the water. On land, the projections act sort of like a cat's whiskers, sending sensory information to special cells under the mole's skin. In water, the projections can detect electrical activity caused by the muscles and nerve activity of potential prey animals. Once the mole captures an unlucky insect or mollusc, the projections also come in handy; the mole can use them to manipulate its food in much the same way that we use our fingers.

Beware the Bait

Most fish are not exactly comely, and this is especially true of the anglerfish. This rather grotesque-looking fish has a basketball-shaped body that may or may not be covered with spines, depending on the species (there are at least 200 known species worldwide). It also has a super-long lower jaw, a short, stumpy upper jaw and a mouth full of really nasty-looking, inwardly angled teeth. Perhaps the strangest feature of this fish, though, is the barbel that projects from its back and grows forward to hang over its face, looking very much like a fishing pole. The barbel, which is actually a long, modified dorsal spine, has a bio-luminescent nub at the end that acts as a lure for potential prey. The anglerfish can even shake the barbel so that the "lure" wriggles about, making it that much more irresistible to other hungry creatures. When the curious, soon-to-be victim swims close to the lure, the anglerfish snatches it with its jaws and swallows it whole.

> **barbel:** a fleshy, thread-like, tactile appendage that usually projects from a fish's snout or mouth (such as the "whiskers" on a catfish) but can also grow from the fish's back (as in the anglerfish's "lure")

Not Exactly Garden Fare

If we were to hold an "oddest-looking creature in the world" contest, the California sea cucumber should definitely be in the running for first prize. This bizarre beast looks more like a vegetable than an animal. As its name suggests, the sea cucumber looks rather cucumber-like, but certainly not like one I'd like to see in my salad anytime soon. With its leathery, brown-and-red-blotched skin and soft, squishy body, this cucumber appears to have gone *way* past its "best before" date. This echinoderm, a relative of the sea urchin, has no obvious

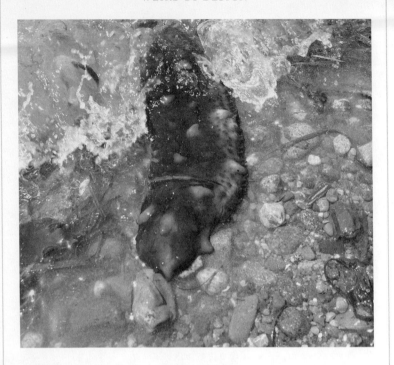

head, so it is hard to tell one end from the other. More than 900 species of sea cucumber exist worldwide, living on the sea floor, basically at all depths. You could think of them as oceanic earthworms; although sea cucumbers and earthworms are not closely related, sea cucumbers serve a similar function in the water as earthworms do on land, breaking down organic matter so that it can be recycled back into the ecosystem.

Filter Feeder

Baleen whales are strange-looking creatures. They range in size from the pygmy right whale, which averages about 4500 kilograms in weight and reaches about 6 metres in length, to the blue whale, which tips (or more likely flattens) the scales at a whopping 200 tonnes and reaches

20 to 30 metres in length—more than half the length of an Olympic swimming pool. Whether big or small, all baleen whales have one thing in common—they are all filter feeders. These whales take their name from the baleen, or whalebone plates, hanging from the top of their mouths. The plates are made of keratin, the same substance that makes up our fingernails.

Baleen whales have no use for teeth, or at least not teeth as we generally think of them. They feed mostly on krill, but also eat plankton and even small fish, all of which are too small to be caught or chewed with regular teeth. So, nature provided modified dentition in the form of the toothbrush-bristle-like plates. As the whale swims, it sucks in huge mouthfuls of water, and then uses its tongue to push the water through the baleen plates, which act like a filter, allowing the water to pass through and trapping the plankton inside the whale's mouth, ready to be swallowed.

There are many species of baleen whale in the world's oceans; these large animals have even larger appetites, so to successfully co-exist, different species have put their own twist on how they feed. Right whales feed at the water's surface, swimming along with their mouths open so surface-dwelling plankton can be scooped up. Rorquals feed underwater by swallowing massive amounts of water; they have throat grooves that expand, and these whales can hold just under 69,000 litres at a time! Grey whales have yet another method of filter feeding; they feed at the bottom of the water column, swimming along the ocean floor on their sides while sucking up sediment and whatever little creatures are hiding in it. The greys then use their baleen to push out the sand and mud, trapping and swallowing the crustaceans.

Raisin of the Sea

Although baleen whales are an odd-looking lot, the sperm whale is even stranger looking, despite its rows of pearly whites. First there is the oddly shaped, square-ended head. Then there is the skin. Sperm whales are covered with wrinkly skin. They look sort of like giant, marine dates or raisins, albeit streamlined ones that are square at one end and have at tail at the other. I can't quite call them the whale equivalent of the Shar pei (Chinese wrinkle dog) because their skin doesn't hang in folds off their frames—it is just really wrinkled. Researchers believe that the wrinkles serve a purpose other than just decoration and may help reduce friction by reducing the amount of turbulence the whale experiences when swimming through the water.

The Better to See You With

Species of the *Histioteuthis* genus of squid are colloquially referred to as cock-eyed squid, and for good reason. These bizarre-looking animals have eyes that are different sizes. The small eye points to the sea bottom, searching for bioluminescence from potential predators or prey, while the larger eye, which is more than twice the size of the small one, points towards the ocean's surface, taking advantage of the light that filters down through the water column. And mismatched eyes are not the only feature that makes these squid a bit less than attractive. Their photophores create a pattern on their bodies that has been likened to the measles. So the freaky-eyed squids also look diseased. Nice. The reverse jewell squid (*Histioteuthis reversa*) is found in Canadian waters, in both the Atlantic and Pacific oceans, and grows to about one metre in length.

> **photophores:** light-producing organs found in certain marine animals

STRUCTURALLY STRANGE

Foldable Fangs

A rattlesnake's fangs fold back into the snake's mouth when not in use. Then, when the snake needs them, it opens its mouth wide, and *voila!* There they are, ready to penetrate the hide of an unsuspecting victim. A rattlesnake that loses a fang—whether it breaks off in the snake's victim (like perhaps the tough boot leather of a hapless hiker) or is extracted by an ignorant pet owner hoping to avoid being on the receiving end of a venomous bite—isn't in such dire straits as one might expect; given a little time, the fang simply grows back. Rattlesnakes replace their teeth about every month, one side at a time. If the old fang is still intact when the new one grows in, the snake will have two teeth on one side of its mouth for a short time. If it loses a fang before the next one is ready to come in, I guess the snake just has to gum the poison into its prey while it waits for the new fang to grow, which by the way, it can do. A rattlesnake can still get poison into its prey even if it has no fangs.

Vicious Venom

Another weird thing about a rattlesnake's fangs is that they are hollow. You'd think being hollow might make the fangs a little less sturdy, but that doesn't seem to be the case. What the hollow fangs do is provide a conduit for venom delivery. Venom, which is stored in glands behind and below the rattlesnake's eyes, is forced up through the hollow fangs and into the flesh of whatever unlucky creature the rattlesnake has sunk its teeth into. The snake can control how much venom it injects by contracting the muscles that surround the venom glands. A rattlesnake's venom not only paralyzes and eventually kills the prey, it also begins breaking down the tissue, making it easier for the snake to digest.

The Better to Smell You With…
Well, Not Really

Have you ever taken a close look at a rattlesnake's face? If you have, you probably noticed that the snake seems to have two sets of nostrils. No, the snake does not have some freaky, multi-nostril respiratory system. The extra "nostrils" are in fact facial pits, heat-sensitive organs that the snake uses to make locating prey a bit easier. With the help of its facial pits, the snake can sense if a warm-blooded potential prey animal is nearby and can zero in on its exact whereabouts.

The pits also help the snake hear better; because they do not have external ears like we do, rattlesnakes must instead sense vibrations, and the pits help them do so. And as if that wasn't enough, the pits help improve the snake's sense of smell. So I guess in a way, the pits *are* like a set of nostrils… super nostrils!

Mystery Goo: Purpose Unknown

Even someone new to the pursuit of whale watching is unlikely to mistake the sperm whale for any other species. With its huge, flat or square-ended head, the sperm whale is in a class all its own. Its distinctive head shape is a result of the large cavity in its forehead (and when I say large, I really mean it—a compact car could fit inside it with room to spare). The cavity is filled with spermaceti, an oily substance that early whalers mistakenly believed to be sperm (hence the whale's common name). No one is quite sure what purpose the spermaceti serves as far as the whale is concerned, but theories abound. It has been suggested that the spermaceti may help the whale regulate its buoyancy, or it might serve a purpose in echolocation. But one thing is certain—spermaceti was the reason behind the slaughter of vast numbers of whales from the mid-18th century until 1985, when the International Whaling Commission afforded the sperm whale full protected status. Spermaceti was a substance much in demand and was used to lubricate small engines, such as those in clocks, as well as to make an array of other products, including candles and soap. The sperm whale has the largest quantity of spermaceti of any whale, so it was a favourite target of whalers.

I Won't Back Down

The raccoon can rotate its hind feet 80 degrees so that it can climb headfirst down a tree trunk instead of the less-graceful butt-first descent. It is one of the only mammals that can do so.

At the Tip of the Tongue

If you were to take a stroll through a wooded area in summer, you would most likely hear the drumming of a woodpecker as it hammers its bill against a tree trunk, trying to get at the tasty insects inside. But working a hole through the tree's tough bark is just the first challenge the hungry bird faces; it still has to catch the little critters as they race about in tunnels they've created in the wood. After all, the insects are hardly going to crawl willingly into the woodpecker's hole, within reach of its beak, now are they? Thankfully for the woodpecker, it doesn't need to rely on its beak to catch a meal; instead, the bird has a super-long tongue that it sticks into the hole in the bark, then works into the tunnels in the wood, impaling or scooping out the bugs. A woodpecker's tongue is much longer than its bill, so to fit in the bird's head, the tongue cannot be attached to the inside of the bird's mouth. Instead, the tongue wraps around the bird's skull. The woodpecker can extend its tongue when necessary, then roll it back up inside its skull, rather like a tape measure inside its case, when it has finished feeding.

In the Eye of the Beholder

Octopus bodies are just plain weird. Amazing, but weird. To support my statement, I offer the following two facts. First, an octopus has three hearts. Yup, that's right. Three. Its blood cannot carry oxygen efficiently, so the octopus needs three hearts to be sure that enough blood gets pumped through its system to meet its oxygen requirements. Not weird enough for you? Okay, how about this? An octopus can squeeze its body through an opening that is the same size as its eyeball. Because an octopus has no exoskeleton, its body is pliable. Well, most of its body is pliable. Its eyeballs are not; they are pretty much fixed in size. So, if the octopus' eyes can make it through the hole or crevice, the rest of the body will follow. Like I said, weird.

Hard to Stomach

It is common knowledge that cows have four stomachs; this fact has been a part of our collective knowledge for so long it has lost its sense of weirdness. But how about the fact that birds, regardless of the species, have two. Even the tiniest hummingbird has two itty-bitty stomachs in its teeny little frame. The first stomach, called the proventriculus, is the glandular stomach. Food passes down the esophagus into the proventriculus, where it mixes with digestive enzymes secreted by glands in the stomach. From there, the food gets pushed into the second stomach, or gizzard, which is the muscular stomach. The gizzard contains pebbles and grit swallowed by the bird to help grind up the food. Because birds do not chew their food, it must be broken down in the gizzard. In some birds, such as owls, the gizzard traps the indigestible bits (bones, beaks, fur) and forms them into pellets that the bird regurgitates.

Never Mind Sticky Fingers,
How About Sticky Feet?

Have you ever watched a treefrog move about? They are
amazing to watch. They can creep up tree trunks, walls,
pretty much any vertical surface with ease, and they can
even hang upside down from smooth surfaces such as wet
leaves. Now it's one thing for humans, with our handy
opposable thumbs, to be able to grip something while hanging
upside down, but the last time I checked, treefrogs were
decidedly thumb-less.

Fortunately for the frogs, their toe pads are coated with
a film of sticky mucus that helps them cling to surfaces by
wet adhesion. However, sometimes wet adhesion isn't enough;

although it allows the frog to stick to rough surfaces, it is not so helpful when the frog wants to grip something smooth. That's where dry friction comes in. The treefrog's toe pads also have microscopic bumps—sort of like the cleats on soccer shoes—that cut through the sticky mucus. These bumps, again like soccer cleats, allow the frog to grip smooth surfaces. And when the frog feels the need to hang upside down, the watery mucus on its toe pads creates surface tension so that its feet stay firmly attached to the underside of whatever it is clinging to. So, no matter what kind of terrain the little guy wants to move across, his fabulous feet have got it covered!

A Frog Tale

The tailed frog nudges the Pacific treefrog out as a contender for smallest-frog-in-Canada status. When full grown, this tiny frog reaches only 2.5 to 4.5 centimetres (versus the treefrog's 5 centimetres) from its nose to the tip of its tail. Yup, that's right, tail. It's not called the tailed frog for nothing. All frogs have a tail in their tadpole form, but the male tailed frog keeps its tail into adulthood—or, more appropriately, a tail-like appendage. Whereas male frogs of all other Canadian (in fact, all other North American) species fertilize the female's eggs externally, coating them with sperm after the female has laid them, tailed frog males fertilize the eggs while the female is still carrying them internally. The male's infamous "tail" is actually the organ he uses to inseminate his mate.

DID YOU KNOW?

In Canada, there are two tailed frog species, both of which are found only in BC. The coastal tailed frog, as its name suggests, is found along the coast, and the Rocky Mountain tailed frog is found in the Kootenays (which makes one wonder why it is not called the Kootenay tailed frog, but I digress).

Late Bloomer

Still on the tailed frog—for such a little creature, it sure takes a long time to metamorphose from tadpole to frog. The cycle from egg to fully developed adult can take up to four years! Even then, it doesn't become sexually mature for another few years. Good thing this frog has such a long lifespan—it can live to be 15 to 20 years old. The average frog lives only about four to eight years. Tailed frogs are the closest living relatives to the first amphibians, which emerged in the Paleozoic era, over 250 million years ago.

Size Matters, All Right

The male walrus has the dubious distinction of being the mammal with the longest penis bone, or baculum, in the animal kingdom. It can reach more than half a metre in length. Now men, before you break into spontaneous applause, read on. Walruses are large animals and are not exactly featherweights. Although they may be graceful in the water, on land, as I'm sure you can imagine, their weight makes them a bit ungainly. Here comes the weird bit. The male, as he hauls himself out of the water and onto an ice floe, or as he drags his bulk over rough or rocky terrain, often breaks his penis bone. Yup, snaps it right in half. Suddenly a long schlong doesn't seem so attractive, does it?

It's Supposed to Look Like That

The shape of the crosswing's beak sort of brings to mind a bird that has flown beak-first into a pane of glass or a wall, and instead of breaking, the bill has been bent so that the top mandible crosses over the lower one. Of course, that is not the case. Actually, the beak's specialized shape is an adaptation that helps the crossbill extract the seeds from its preferred food source, pine cones. To get at the seeds, the bird wedges its beak between the scales of the cone and pries them apart. It can then use its tongue to scoop out the seed. Birds that do not have the crossbill's specialized beak can still feed on conifer seeds, but not with the crossbill's speed and gusto. Thanks to its beak, the crossbill can eat up to 3000 conifer seeds in a day; however, the specialized bill is obviously a disadvantage when the bird tries to feed on other foods.

Young crossbills are born with straight bills that gradually cross as the birds age. Scientists do not know what causes the mandibles to cross or what determines the direction in which they cross. The ratio of left and right crossings is about equal.

In Canada, we have two species of crossbills: the red crossbill and the white-winged crossbill. These birds can be found in the boreal forest throughout Canada.

Mmm…Tasty!

When most people think of an animal "hearing" or "listening," they think of the animal using its ears, but the Cuvier's beaked whale uses its jaw and throat. Toothed whales have a structure, called the "acoustic window," in the lower jaw through which sound enters the whale's head, or so it was believed. The "window" is basically a thick layer of fat covered by a thin layer of bone; scientists had previously thought that sound waves made the bone vibrate and then travelled through the layer of fat into the whale's ear. However, this doesn't seem to be the case for Cuvier's beaked whale. Instead, sound waves travel under the animal's jaw, passing through the throat before entering a hole at the back of the jaw to reach the layer of fat beside the ear. So, basically, the whale swallows sound to hear!

Cuvier's beaked whales live in all oceans of the world, except in polar regions. They are one of the least studied whales, rarely seen unless they are beached, and not much is known about their life history. However, these whales get stranded more often than any other species of beaked whale, and it has been suggested that the military's use of sonar interferes with the whales' echolocation, leading them astray. Scientists hope that by understanding the mechanisms by which these whales hear, they will also be able to determine what effect, if any, the use of sonar in submarines has on these animals.

FLAWED DESIGN

Just a Little Off the Top

The magnificent frigatebird is a seabird. It spends a great deal of its time flying over the open ocean, searching for food. When it sees something edible, such as a fish, it swoops down and snatches the creature from the water's surface or just below it. But there is one little flaw in this bird's design—its feathers are not waterproof. If the frigatebird were to land on the water instead of just skimming over the top of it, its feathers would become waterlogged and it would most likely drown.

Nitrogen Overload

You'd think that, as the champion of deep-sea diving, the sperm whale would have the necessary body adaptations to prevent it from being susceptible to the bends, otherwise known as decompression sickness. Not so. As any scuba diver knows, decompression sickness occurs when one is resurfacing from a deep dive because the change in pressure causes nitrogen in the body to come out of solution and form bubbles in body tissue and the bloodstream. In humans, symptoms range from mild pain in the joints or skin rashes to numbness, dizziness and, in extreme cases, paralysis or even death. In sperm whales, apparently the bends are manifested not by skin rashes or death, but by pitting and erosion of the nose and ribs.

Too Much of a Good Thing

The bodies of grasshoppers and butterflies cannot regulate the amount of oxygen that reaches their muscles. Mammals depend on proteins such as hemoglobin to carry oxygen in the blood, which then limits how much oxygen our bodies hold. Grasshoppers and butterflies, however, do not have such proteins. Instead, the oxygen they breathe goes directly to their tissues through a branching respiratory system. Too much oxygen in the body can be toxic. Therefore, to avoid poisoning themselves with an oxygen overload, the critters must stop breathing for a while. To do this, they close special openings on their bodies, called spiracles, which work like valves, stopping the flow of oxygen into their tissues.

FORM FOLLOWS FUNCTION

Frogsicle, Anyone?

As the most widely distributed frog above the Arctic Circle, the wood frog has obviously honed its cold-weather survival skills. It spends the winter sheltering under leaf litter, where it does not have the protection of an insulating layer of mud or dirt; basically, it is at the mercy of the elements. However, this little amphibian doesn't bother wasting its energy trying to protect itself from the harsh cold; instead, as the landscape freezes under the cloak of winter, the frog freezes right along with it.

The wood frog has special proteins in its blood, called nucleating proteins, that allow water in the blood to freeze first, so that the resulting ice can draw the water out of the frog's cells. As the cell water is drawn out, it is replaced by glucose, which is produced in the liver. The glucose in the cell allows the cell to keep its shape rather than collapsing with the loss of water. As I'm sure you remember from your grade school science class, when water freezes, it expands and crystallizes. Water that freezes inside a cell bursts or tears the cell wall as it freezes, but because the frozen frog has no water in its cells, they do not freeze and are therefore not damaged by ice.

While the frog is frozen, its internal organs are encased in ice, and the space in its abdominal cavity is filled with a solid chunk of ice. The frog does not breathe, its heart does not beat and its brain shows no activity. Its brain also receives no oxygen, which would cause brain damage in a human but seems to have no negative impact on the frog. When spring comes and the frog begins to thaw, the ice in its body melts, and water seeps back into the cells. Once the heart thaws enough to start beating again, it pumps blood throughout the frog's body, and the frog revives enough to go about its business. The wood frog's natural history is not well enough understood to know if freezing affects its lifespan, but researchers have determined that male frogs that have been frozen are less active, less vocal and have more trouble identifying potential mates than individuals that have not been frozen.

Bustin' Loose

I confess that until recently I'd never really given much thought to the whole bird reproduction process, but when you really stop to think about it, hatching from an egg is really quite an extraordinary way to come into this world. When I pictured a chick encased in its egg, I suppose I kind of saw it as a separate entity nestled in the middle of the

egg space, protected by the shell but not a part of it, sort of like a little downy (or bald, depending on the species) version of the prize in one of those chocolate eggs with the toys inside. In my mind, when the chick was ready for a little taste of freedom, it would simply crack its egg and casually step over the broken shell fragments into the great world beyond. Clearly, I did not spend enough time on a farm as a child.

In truth, the process is not quite so simple. The chick is actually attached to the inside of the eggshell via a system of membranes. Once the chick uses its egg tooth (a small, sharp, horn-like growth on its beak that falls off soon after the chick hatches) to chip a hole in the shell, fresh air rushes into the egg, giving the chick its first breath, but also drying the membranes that attach it to the shell. As the membranes dry, blood vessels within them begin to shrink, forcing the blood to run into the chick's body. From here on, it is all a matter of timing. If the membranes are torn too soon, the chick will bleed to death. If the chick doesn't get out of the egg before the membranes dry out completely, it will be trapped inside and, again, will die. See, not so simple after all. Extraordinary, really.

Hot Stuff

Cold waters don't seem to faze the salmon shark one bit; it can be found throughout the waters of the northern Pacific Ocean, including offshore along the coast of BC and even reaching as far north as Alaska. The salmon shark is in constant motion—it has to be, otherwise it could become hypothermic. This shark is endothermic and has the highest body temperature of any shark species. Because it is in constant motion, it generates enough body heat to keep it warmer than the surrounding water—up to 20°C warmer, depending on where and how deep it is swimming. Regardless of its environs, a salmon shark's core temperature usually stays around 25°C. The fish has a special vascular network, called the "rete mirabile," that protects its eyes, brain and muscle tissue from damage associated with temperature changes. The rete mirabile is basically a biological heat-exchange system. Within the rete, there are two blood-streams that lay close to each other, with the blood in each system flowing in the opposite direction. In this way, the warmer blood in the arteries, coming from the shark's core, passes by blood in the veins that is coming from the shark's extremities and has been cooled by the icy seawater. Heat from the warmer blood gets passed to the colder blood as it flows by, keeping the shark's overall body temperature pretty stable and warmer than it would otherwise be. Because the salmon shark runs warm, it is also one of the world's fastest sharks. To fuel its speedy travel, the shark stores energy in its liver, which grows so large that it fills most of the shark's body cavity.

> **endothermic:** an organism that is able to produce its own body heat and maintain a constant temperature

Always on the Wing

The chimney swift spends a great deal of time in the air. Obviously, it must land to lay its eggs and rear its young, and it also roosts in chimneys overnight (hence its name), but that is pretty much the only time that it gives its wings a rest. Bathing, eating, drinking and sometimes even mating all take place while the bird is in flight. Because it spends so little time perched, this species doesn't really have a lot of use for its legs. Consequently, they have all but withered away. The swift obviously still has legs, but they are short and stumpy and are not strong enough to support its weight should it land on a branch or the ground. Instead, the bird uses its feet to cling to vertical surfaces such as walls or the insides of chimneys.

Eating Light

The muskox lives in one of the most inhospitable environments in the world—the Canadian Arctic. Much of the year, the animal is buffeted by harsh, icy winds, and food is not exactly bountiful. Its long outer coat and thick, insulating inner hair help protect the muskox from suffering the cold, but they don't help stave off starvation. Thankfully, the muskox has another handy little adaptation to help on that front—it can lower its metabolism by about 20 percent. By lowering its metabolism, it reduces its energy consumption, which in turn reduces how much food it needs to eat to stay alive during the harsh winter months. The muskox is believed to have come to what is now the Canadian north by crossing the Bering Strait about 90,000 years ago. It has been around since the time of the mammoths and even shared the frozen landscape with these long-gone creatures.

You've Got a Certain Glow...

When a living animal or plant gives off its own light, that light is called bioluminescence. Although it does occur in some surface creatures such as glow-worms and fireflies, it is much more prevalent in marine animals; almost 70 percent of creatures living in deep ocean waters have some form of bioluminescence. I guess when you live in an area of perpetual darkness, the only way you are going to get any light is if you generate it yourself.

Most bioluminescent animals produce their light in specially designed organs called photophores. Although they contain gland-like cells that are the actual source of the light, not all photophores are created equal. Some are a bit fancier in design and have special adaptations such as flaps of skin that can be used to turn the light on or off, special filters that allow the light to be focused in a specific direction, and even colour filters. Bioluminescent animals that do not have photophores get their light from certain light-producing bacteria; in this case, the animal and the bacteria have a symbiotic relationship.

Bioluminescence can be put to a number of good uses. Flashing your bio-light at a predator is a pretty good way to startle and perhaps distract it. It is also a good way to blind the predator while you beat a hasty retreat. The positioning of the photophores can also confuse a predator so that it isn't sure which end of its prey it should be attacking.

Some creatures do not emit light themselves, but instead squirt a luminescent cloud into the water, much like an octopus does with its ink. For a predator, bioluminescence can light up the surrounding waters so the creature can actually see its prey, or a more retiring predator can use it to lure potential prey close enough to be snatched and swallowed. Animals can also use their bioluminescence to communicate with other members of their species or try to attract a mate.

To Breathe or Not to Breathe

For land-based mammals, nothing comes more naturally than breathing—we don't have to think about it, we just do it. In fact, even if we try not to take a breath, our body does it for us anyway. But marine mammals don't have it quite so easy. They must decide when they are going to draw a breath because they have to be sure that they are in a position to do so—that is, at the water's surface. This means that they must consciously decide when to breathe—the keyword here being "conscious." If the whale has to be conscious to breathe, and it needs to breathe every half-hour or so, when does it get to sleep? Even a whale needs its rest, right?

Whales do not sleep the same way humans do; rather than the total unconsciousness that characterizes human slumber, whales sleep in a semiconscious state. They shut down one half of their brain at a time, letting the sleeping side get the rest is needs, while the awake side stays alert enough to allow the whale to carry on doing whatever it needs to do to survive. Researchers have likened it to the drifting, semi-aware state that we slip into as we fall asleep.

Sleeping on the Job

Whales are not the only animals that sleep one brain hemisphere at a time. Some species of birds, such as mallards, also rest one lobe while the other stays alert. Sleepy birds often group together into a loose circle, with a number of birds in the middle surrounded by a ring of individuals that position themselves in such a way that one eye faces the inner circle and one eye faces outwards. The outward-facing eye is the alert hemisphere of the brain, and it keeps a lookout for potential predators, while the hemisphere associated with the inner-facing eye shuts down and gets some rest. The birds shift position from time to time, turning around so they can rest the other side of their brain as well. Birds in the inner circle rest both lobes simultaneously and often move to the outer circle after they've had enough sleep, letting the sentry birds leave their posts and get some full-brain shuteye.

Catching a Few Zs

Walruses spend a great deal of their time in the water—two-thirds of their lives, in fact. They feed, court prospective partners, mate, give birth and even nurse their young while in the water. So it should be no surprise, then, that they also like to catch a little shuteye while at sea. When they sleep, walruses rest both hemispheres of their brain at the same time, like we do, not one half at a time, as do whales. So how does a sleeping walrus stay afloat instead of dropping like a particularly heavy stone to the bottom of the ocean? It inflates its pharyngeal pouches—large sacs on either side of its esophagus that can hold up to 50 litres of air. When fully inflated, these pouches allow the walrus to bob like a cork in the water, floating high enough so that its head

effortlessly stays above the ocean's surface. To prevent itself from drifting along at the mercy of the ocean's currents, a sleepy walrus uses its tusks to anchor itself to a stable ice floe.

Faking It

Birds in the corvid family—blue jays, ravens and crows—are some of the most intelligent in the avian kingdom. They are known for their boldness and their cleverness, and one species, the blue jay, is also known for its powers of imitation. The blue jay can do a stunningly accurate rendition of a red-tailed hawk call and has also been observed imitating the calls of ospreys, American kestrels, Cooper's hawks, eastern screech-owls and even their close relatives, the crows, among others. Researchers have offered several different theories as to why the jay imitates other birds' calls. Some have suggested that the jay imitates a hawk to warn other jays that the bird of prey is nearby; another theory is that the jay imitates the hawk to trick other birds in the area into thinking that they are in danger.

It's easy to see how this little trick could come in handy. Say, for example, that a jay wants to visit a particularly busy bird feeder but can't be bothered to muscle its way through the hoards of little songbirds helping themselves to the seeds. If the jay let out a convincing hawk call, the songbirds would abandon the feeder and dive for cover, and the jay could then feed at its leisure. However, neither of the theories mentioned above explains why the bird also imitates non-avian sounds such as cat meows, alarm clocks, cell-phone ring tones and human words and whistles. Perhaps the clever birds are just toying with us and having a laugh at our expense.

Deep-Sea Champion

The sperm whale is one of deepest divers (if not the deepest) in the ocean. This whale can dive down to more than 3000 metres (though 1000-or-so-metre dives are more common) and can stay underwater for up to 90 minutes while hunting for its favourite food, squid. Its body has a few strange features that allow the whale to dive to such great depths. First, myoglobin (a type of hemoglobin) in the whale's muscles stores oxygen, which it then releases back to the muscles and vital organs as needed while the whale is underwater. In this way, the muscles and other vital bits get the oxygen they need to keep functioning, even though the whale cannot draw a breath. Also, the lungs shrink, condensing the whale's last breath into a smaller space so that it is about one-quarter of the volume it was at the ocean's surface.

Furthermore, during deep dives, the whale's rib cage "folds up" and collapses. A human's rib cage would also collapse at the depth these whales frequent because of the tremendous amount of pressure from the surrounding water, but we would not survive the experience. Fortunately for the whale, its body is designed in such a way that the collapsed rib cage regains its uncrumpled state once the whale resurfaces.

> **myoglobin:** an iron-containing protein that carries and stores oxygen in muscle cells

Toad in the Hole

The plains spadefoot toad also lives in the hot, arid regions of the southern Prairies, which is not really a great place to call home if you are an amphibian and need to keep your skin moist to survive. But natural selection being what it is, the toad has a few handy adaptations that allow it to get by in its desert home. First, it spends a great deal of its life under the sand, keeping cool. This toad has a spur (or "spade") on the bottom of each hind foot (hence the name "spadefoot") that it uses to bury itself in the sand. To achieve this feat, the toad digs in backwards (that is, butt first), using the spurs on its hind feet to scoop out the sand. It digs through the drier upper layers of soil until it reaches the moister soil underneath, then it remains there, only venturing out to feed once the sun has gone down. If its soil layer gets a little too dry for comfort, another of the spadefoot's neat adaptations comes into play—the toad will make a sort of "cocoon" out of its shed skin, which helps to keep moisture next to its body. During periods of extreme, prolonged heat, the spadefoot goes into estivation, a sort of summer hibernation in which the toad does not need to eat or drink for extended periods of time.

Water Conservation

The Ord's kangaroo rat makes its home in the arid regions of southern Saskatchewan and Alberta. Water is not easy to come by in its habitat, so the kangaroo rat has adapted to doing without. It doesn't take in any moisture by drinking, and its diet consists mostly of dry foods, such as grains and seeds. Scientists who have studied this rat's eating habits estimate that the little creature gets only about one-tenth of the water a mammal of its size requires to survive. So how does the little guy do it? Well, its body creates and stores its own water, called "metabolic water," as food is digested. Also, the kangaroo rat's digestive system is so efficient that its body loses little water in expelling waste—the rat's urine is four times more concentrated than that of humans, and its stool is almost bone dry. Most mammals lose a lot of moisture through respiration, but the kangaroo rat's ultra-efficient body has got this covered as well; when the animal exhales, its cool nose condenses the water vapour from its lungs before it is breathed out.

Sleeping It Off

Nobody likes to venture out in nasty weather, not even chimney swifts or, more importantly, the flying insects that they feed on. So when it is miserable outside and there are few, if any, flying insects to be seen, the swifts often have to go without food. Fortunately for the birds, they can slip into a state of torpor, sort of like hibernation in mammals, to conserve energy. When a bird is in this state, its body temperature drops, it does not react to outside stimulation, and its breathing slows dramatically. Because it requires less oxygen, it does not draw heavily on its fat reserves. While in torpor, both young and adult swifts can survive for about 10 days without eating. The chicks can lose up to 50 percent of their body weight and be no worse off for the experience.

> **torpor**: a state of inactivity or decreased activity in an animal, rather like a mild form of hibernation, during which metabolism slows and body temperature drops

A Pink Walrus Is a Warm Walrus

Quick, without thinking about it, tell me what colour a walrus is. Did you say grey? Brownish? I'm betting you didn't say pinkish, but you could have, and you would still have been correct. As an adaptation for keeping warm in near-freezing Arctic water, the walrus' blood vessels constrict in the cold water, moving the blood away from the skin, where it would quickly be cooled by the surrounding frigid water, and towards the animal's main organs, where the body's core temperature will keep it warmer. So, when the walrus is cold, its skin takes on a whitish appearance, and when the walrus is toasty warm, the skin has a definite pink tone to it.

Bear Bones

If you or I were to lay around for months at a time, we would experience a debilitating loss of bone mass. When we are inactive, our bone regeneration either slows or stops entirely, leaving our bones brittle. Not so with bears. They sleep away the winter months tucked cozily in their dens and wake in the spring a little thinner in girth, but no worse for wear, bone-wise. No one knows how a bear's body can keep regenerating bone even when the bear is lolling about, but it is most likely related to a hormone or chemical in the bear's body. Researchers hope that by unravelling the mystery of what protects the bear from losing its bone mass, they might eventually be able to apply the knowledge to helping people who suffer from osteoporosis.

Talk About Foul Breath

Deep-sea creatures are some of the most mysterious life forms on the planet. A good example would be the sea cucumber. When you are a headless, limbless creature plodding along the ocean floor in perpetual darkness, with tremendous pressure constantly bearing down on you from the water above, I guess the biological norms seen in terrestrial animals really shouldn't apply. This probably explains why the sea cucumber breathes through its anus. Better it than me…

CHEMICAL

Toxic Flatulence

Many animals have unique adaptations that help keep predators at bay, but in my mind, none are quite as spectacular as that of the bombardier beetle. There are many species of bombardier beetle; North America alone has almost 50 species, including the American bombardier beetle, *Brachinus fumans,* which can be found in southern Ontario. So what makes this beetle so unique? Well, it shoots a hot chemical spray—as hot as boiling water—from its butt. The beetle has two glands at the rear of its abdomen, each of which has two chambers. The inner chamber contains hydrogen peroxide and hydroquinones, and the outer chamber holds catalase and peroxidase. When the beetle feels threatened, it opens a valve that usually keeps the chemicals in the two chambers separate. As the chemicals in the inner chamber pass through the outer chamber, they mix with those chemicals, creating a sort of bomb, which is then explosively expelled out the beetle's backside towards a would-be attacker. The explosion has so much force that it actually makes an audible bang. For maximum effect, the beetle releases the chemicals in pulses rather than a steady stream, and it can rotate its abdomen 270 degrees to aim its explosive force. It can also shoot the chemicals over its back, where a pair of reflectors ricochets the spray into the direction the beetle wants it to go.

catalase: an enzyme found in plant and animal cells that decomposes hydrogen into oxygen and water

hydroquinones: aromatic organic compounds that are a type of phenol; one of the primary reagents in the defensive glands of the bombardier beetle

peroxidase: an enzyme that speeds up the break down of hydrogen peroxide into oxygen and water

Now You See Me...

The octopus and squid are not the only sea creatures that squirt a screen of ink into the water to cloak their movements; the pygmy sperm whale, a much smaller relative of the sperm whale that is found in temperate waters worldwide, squirts a reddish substance from its intestines into the water when it is startled or if it feels threatened. This substance makes the water murky so that the frightened little guy can make its escape. You don't want to swim too close to this whale; you might get a face full of something nasty. At least with the octopus and squid, you know the mess they eject is ink. Who knows what might come shooting out of this guy's intestines?

"Fouler's" Toad?

This toad seems to have taken a little bit of everything from the defensive strategy armoury. It has at its disposal camouflage, the use of toxins and mimicry—talk about defensive overkill! With a body length of about 5 to 7 centimetres, the Fowler's toad probably looks like an appetizing and rather substantial snack to many small predators, but this toad is no easy prey. For a time, scientists could not agree whether the toad was nocturnal or diurnal, but as it turns out, the answer is both! As an adult, it is most active at night, but as a juvenile, it moves about mostly during the day. Many amphibians have nocturnal lifestyles to reduce the risk of being eaten by a predator, and it seems to work for this toad—adults of the species have a much higher survival rate than juveniles. However, a Fowler's toad that ventures out when the sun is still up also has a few survival tricks that give it a fighting chance.

As its first line of defence, the amphibian has its cryptic colouration, which allows it to blend into its surroundings and hopefully go unnoticed by predators. Should a particularly keen-eyed hunter cotton on to its presence, the toad can fall back on Plan B—playing dead. If this isn't enough to persuade the predator to walk on by, the toad still has a final defensive adaptation—the warts on its body contain a toxic secretion that oozes from the skin. If an animal tries to bite the toad, it gets a mouthful of toxin that will, at the very least, irritate the sensitive skin in its mouth and, in a worst-case scenario, can kill the animal outright.

Despite it arsenal of defensive adaptations, this toad is struggling in Canada, where it can be found in southern Ontario, the northern limit of its range. It has been protected under the Fish and Wildlife Act since 1997, and in 1999, it was added to the Committee on the Status of Endangered Wildlife in Canada (COSEWIC) list of threatened species, where it remains today.

Getting Skunked

Being on the receiving end of a mist or stream of skunk spray probably ranks pretty high on most people's things-I'd-prefer-not-to-experience-in-my-lifetime list, but really, the recipient, or sprayee, if you prefer, has only him- or herself to blame should such an event occur.

Skunks do not just run about, haphazardly spraying their noxious fumes; they give clear warnings when they are not pleased, trying to let you, their potential target, know that you will not like what is coming if you don't back off. First the skunk will face you head on and stamp its little front paws, hissing and growling; then, if that isn't intimidating enough to send you on your way, the agitated skunk lifts its tail—a clear indication that something nasty is in the works. If you are foolish enough to stick around after that, the skunk changes its body's orientation, pointing its butt

in your direction instead of facing you directly, but twisting its body into a "C" shape so that it can still keep a wary eye on you as well (or maybe it just wants to see your reaction once it lets the smelly stuff fly). 'Cause that's what happens next—if you are anywhere within 3 metres of the agitated skunk, you will be doused with some of the foulest-smelling liquid you will ever encounter. The reeking musk comes from two glands, one on either side of the skunk's anus, and it passes through a nipple that the animal can aim with amazing accuracy, controlling where it directs the smelly flow. A skunk can also choose between sending out a fine, blanketing mist of skunk juice or shooting out a more concentrated spray, sort of like the nozzle on one of those spray bottles people use for spritzing their plants. Generally, the skunk aims for its opponent's face or eyes, and any creature unlucky enough to receive a direct hit in the eyes will be temporarily blinded. The musk also burns skin on contact, and it has such an overpoweringly vile smell that it causes nausea or vomiting in most animals that get a face full. And the smell lingers—just ask anyone who has a dog that's been sprayed. I'm talking weeks here, not days. Even the babies are little stinkers; they may look cute and cuddly, but be warned—by the time they are three weeks old, they can spray just as nastily as their parents.

The good news is that a skunk prefers not to use its smelly defence unless absolutely necessary. Once the sulphury musk has been sprayed, the skunk's body replaces it rather slowly, only one or two teaspoonfuls per week. Without its chemical weapon, the little creature is essentially defenceless—it is not exactly known for being fleet of foot or a particularly good fighter, and with its bold black-and-white colouration, it doesn't exactly blend into the surrounding shrubbery.

Volatile Vomit

Okay, here's the scenario. A little chick sits unprotected in its nest. The parents have gone to sea, hunting for fish or other tasty morsels to feed their baby. Suddenly, a hungry gull swoops in, intent on making a snack of the little nestling. What's a little chick to do? Well, if it is a fulmar chick, it could shoot a load of well-aimed projectile vomit at the intruder, preferably hitting it in the face. Fulmars produce a yellowish, sweet-fishy-smelling oil in their stomachs, made up partly from the oil that occurs in the fish or other foods they eat. Once the oily vomit has been expelled from the chick's stomach, it is cooled by the surrounding air and begins to solidify into a nasty-smelling wax.

Because fulmar chicks spend so much of their time alone in the nest while the parents are off fishing, they have to be able to defend themselves, and fast. It only takes a few seconds for a gull or other predacious bird to swoop in and snatch the chick. Fulmar chicks are born with the spitting behaviour ingrained; they do not learn it from their parents. Chicks that have not even completely hatched yet can already spit small amounts of the smelly oil. By the time they are four days old, the chicks can shoot their spit almost half a metre, and as they grow, so does the distance they can projectile vomit. Eventually, the chicks will be able to hit anything within 1.5 metres.

A fulmar chick will indiscriminately spray vomit on any-thing that gets too close, including its own parents as they return to the nest from a hard day's fishing. Non-fulmar birds on the receiving end of a blast of fulmar vomit can actually die from it. The oil clings to the birds' feathers so that they become matted, destroying their waterproofing and insulating properties, which, for a seabird, spells death by drowning. Fulmar adults seen to be immune to the negative properties of the oil, though, and even use it when preening, working small amounts through their body feathers.

DID YOU KNOW?

Fulmars seem to be smelly birds overall. Not only do they have repulsive-smelling stomach oil, but their bodies give off a strong, musky odour as well. Even the eggs have a characteristic smell. It has been suggested that this funky scent is another form of defence, making the birds, and especially the vulnerable eggs, seem less appetizing to would-be predators.

OTHER BODY ADAPTATIONS

Super Goo

Native to the northern Atlantic from Greenland to Canada and the U.S., the hagfish can grow up to one metre long and looks sort of like an eel, only without the fins, jaws or scales. It lives on the sea bottom, where it is a both a scavenger and a predator. The hagfish's rather disgusting hunting strategy involves crawling inside its prey of choice—usually through the animal's mouth, but sometimes through its gills or anus instead—and using its toothed tongue to rip the prey to bits from the inside out.

Although the hagfish has predators of its own, it also has a very effective, and unique, way of protecting itself from those predators. When it feels threatened, the hagfish produces a slimy substance that oozes from glands located along its sides. The slime, which has been described as "mucous goo" by scientists, swells when it touches the seawater until the hagfish is encased in a slippery cocoon. The same glands also secrete thread-like fibres that reinforce the slime so that it is stretchy but strong. Any animal that gets mixed up in the fibre/goo combo becomes trapped and suffocates, so it could make a handy snack for the hagfish, as long as it doesn't mind ingesting a little of its own slime. To make sure it doesn't get trapped in the goo, the hagfish ties itself into a knot, which it moves down the length of its body, effectively squeezing itself free of the goopy mess.

A Prickly Defence Strategy

Despite what you may have been told when you were a kid, porcupines cannot shoot their quills, but that doesn't make the quills any less impressive as a form of defence. The porcupine, also known rather unfortunately as the "quill pig," has about 30,000 quills covering its body. The only places that are unprotected are its underside, including its tail, and its face (there are some quills on the cheeks, but they are pretty small). The quills are basically thick, stiff hairs that are made of keratin, just like human hair, and they range from about 30 millimetres to 13 centimetres long, depending where they are situated on the body. The longest quills are on the animal's rump.

A frightened porcupine will raise its quills, turn its back so that the rump quills are pointed towards its aggressor, stamp its feet and lash its tail back and forth. If the aggressor doesn't back off, it will most likely get a face full of quills from a well-placed slap of the porcupine's tail. The spines detach easily from the porcupine's body, but not so easily from the animal they are stuck into. Each quill has hundreds of tiny barbs at the end that, when stuck in a victim, work the quill deeper into the poor animal's flesh. The quills can even work their way through muscle until they pierce a vital organ and the animal dies—not a very pleasant way to go.

Barbed Babies

Ever wonder how a female porcupine gives birth? Baby porcupines are born with a full set of quills, but thankfully for the mother, the quills are soft for the first few hours of the baby's life. Before long, though, they harden and become the impressive, prickly armour seen on the adults. From the time they are born, baby porcupines display the same defensive behaviour as their parents, raising their quills and even lashing their little tails when they feel threatened.

Inside Out

How desperate does a creature have to be before it sacrifices its internal organs to an attacker? For the sea cucumber, the answer is, not very. This bizarre creature has a special adaptation, called self-evisceration, that allows it to expel its innards out its butt when it feels threatened, then wander off and resume its day-to-day activities while its body regrows the missing organs. Even stranger, in what has been termed the "cuke nuke" by researchers, the ejected matter poisons the surrounding water and kills any creatures unlucky enough to be in the vicinity.

A Lizard Tale

All right, so a red-backed salamander is on the run, being chased by a hungry shrew, and the shrew is closing in. What does the salamander do? It sheds its tail. The salamander's body is constructed in such a way that is has zones of weakness throughout its tail. To lose the tail, the salamander contracts a muscle under its vent and fractures a vertebra at one of the weak zones. Muscles in the stump contract to close the artery in the tail so that the salamander will not bleed out. The bit that has fallen off continues to wriggle around on the

ground, distracting the predator while the salamander makes its getaway. When the new tail grows it, it is made of cartilage rather than bone, and it is usually a slightly different colour than the tail that was shed.

vent: the anus of a lower vertebrate that serves for excretion and reproduction

Boxed In

Simply put, a turtle's shell is made up of its backbone and rib cage, which have been modified to form a protective shell around the animal's soft body. To protect their soft bits, most turtles withdraw their vulnerable head and limbs into the shell. But the box turtle goes one better. Once it has withdrawn its head and legs, this turtle completely closes its shell. The specialized shell has a hinge running across the plastron (the bottom part of the shell) that allows it to be drawn so tightly closed that not even a knife blade can be slipped in.

CAMOUFLAGE AND MIMICRY

Nobody Likes a Copycat

The bold colours of the monarch's wings announce to the world that this butterfly might not be the best choice for an afternoon snack. Monarchs feed on milkweed plants, and as they eat the plants' leaves, they also consume the plants' toxins. Although these toxins do not affect the monarch itself, they do nasty things to any creature that tries to eat it. Therefore, not many creatures are foolish enough to try. Another butterfly, the viceroy, takes advantage of this fact. The bold colours of its wings, which are almost exactly the same as those of the monarch, also announce to the world that is it toxic—only it isn't. The viceroy does not feed on toxic plants and therefore does not carry poisons in its body. It simply capitalizes on the aversion most creatures have to eating the similarly coloured monarch. Sneaky butterfly...

Chameleon Frog

We've all heard of lizards that can change colour to blend into their surroundings, but how about a colour-changing frog? The grey treefrog—native to Ontario, Manitoba and New Brunswick—can do just that. Depending on the colour of the surface it is sitting on, the treefrog can change from nearly black to grey or green and even to off-white. It can also do a mottled combination of colours. However, the tree-frog's ability to change is a gradual process. Unlike the cha-meleon, it does not change to match each surface that it sits on; rather, it adjusts to the overall environment that it inhabits, so a frog living in lush vegetation will be green, whereas a frog living in a drier region will be browner.

Colour Camouflage

Everyone knows that a panicked octopus will squirt purple-black ink into the water to mask its escape and confuse a potential predator. But did you know that an octopus can actually change colour to blend in with its surroundings? It's true. An octopus has special pigment cells, called chromato-phores, in its skin that contain three separate sacs or pouches, each of which holds a different colour pigment. When the animal is stressed, the chromatophores release each pigment individually until the right blend of colour is achieved. And the tricky cephalopod goes one better—it can even change the texture of its skin to match that of its surroundings, making it that much harder to see.

> **cephalopod:** any mollusc of the class Cephalopoda, a group of marine invertebrate animals that includes the octopus, squid and chambered nautilus

DID YOU KNOW?

An octopus will also change colour according to its mood. If you see a white octopus, you know that the creature is scared. If it is red, watch out! That's one angry cephalopod!

Nothing Here But Us Reeds

As do many animals, the American bittern relies on its cryptic colouration to give it a survival advantage. With its light brown back, long, yellowish bill and streaky, brown-and-white breast, this heron-like bird blends in perfectly with the surrounding vegetation in its preferred habitat, freshwater marshes or wet meadows, especially ones containing cattails and bulrushes. The bittern's colouration allows it to pass unnoticed through the reeds as it sneaks up on its chosen prey, and it also helps the bird evade the eyes of predators. In fact, the bittern is so confident in its ability to blend into the background that it prefers to stay put and wait for a predator to pass it by rather

than fly out of the area to safety. But as it waits out the predator, the bittern is not content to hunker down and cower out of sight. No, not this bird. When it feels threatened, the bittern stretches its body out, extending its neck, tipping its head back and pointing its slender bill skyward. It then starts to sway gently from side to side, mimicking the surrounding reeds and other vegetation as they rustle in the breeze. Its eyes are low on its head so that as it sways, it can peer at the threat. If the predator is not convinced that the bittern is just another stalk of vegetation, the bird takes flight and heads to a safer location, often uttering a harsh, barking call as it departs, almost as though expressing its disgust or cussing the interloper for making it take to the air.

The bittern's defensive strategy doesn't always serve the bird well, however. It does not seem to realize that it can only pass for a reed if there are actually other reeds in the area. The bird will adopt the same pose regardless of its surroundings, even if it is in the middle of an open field or on a stretch of road where there is not a single reed to be seen.

The Twig Defence

The inchworm is a favourite of children everywhere because of the distinctive way that it moves along a branch or other surface, first stretching out flat and grabbing the surface with its front legs, then arching its back and pulling its rear end forward. But even more impressive is this little caterpillar's powers of mimicry. When it senses danger, the inchworm lifts its front end off the branch so that it is standing up straight, and then it stays perfectly still, pretending to be just another twig jutting off the main branch. It can hold this position for hours, until it is sure that all signs of danger have passed.

Try Not to Stick Out

A similar strategy is practiced by the stickbug, or stick insect, found in Ontario. Like the inchworm, the stickbug mimics a twig to avoid being seen and eaten by predators. The stickbug's performance is especially convincing because, as its name suggests, it actually looks like a stick. With its long, brown, woody-looking body and long, spindly legs, the bug looks remarkably like a small branch complete with twigs jutting out at different angles. The bug will even sway gently back and forth to simulate a branch moving in a breeze. If for some reason the stickbug loses its grip on the branch it is clinging to and falls to the ground, it lies still until nightfall (nothing to see here but a bit of dead wood!), when it can creep back to its plant under the cover of darkness.

Eyes in the Back of My Head

Because of its relatively small size, there are quite a few predators that would happily consider taking on a northern pygmy owl as a potential snack. To protect itself, the small owl leads a secretive lifestyle, roosting in thickets where it will not catch a predator's eye. In its hiding spot, the bird makes itself look thinner and closes its eyes into small slits, hoping to remain undetected, but this isn't always enough to keep it safe. So, when challenged by a larger bird or mammal, the owl puffs up its feathers and spreads its tail to make itself look more impressive. But the pygmy owl's strangest deception is the pair of false eyespots on the plumage at the back of its head. With this second set of "eyes," the owl dissuades predators from attacking it from behind by tricking them into believing it sees them coming.

BEHAVIOUR

Death Becomes It

Although it may be best known for its feigned-death survival strategy, the scrappy little Virginia opossum first tries to scare a potential predator away. When under attack, the feisty marsupial hisses and snarls, baring its teeth menacingly. If the would-be attacker is unmoved by the display, it is time for Plan B. The opossum suddenly keels over and lies still. I must confess, I don't really understand how this dissuades a predator from attacking. Doesn't it wonder at the abrupt change in the opossum? Doesn't it remember that the same little creature was, just a few seconds ago, hissing and snarling in its face? I suppose the answer must be in the quality of the opossum's performance. Spread out on its side, the pseudo-corpse is dedicated to its act. It keeps its eyes half open but with an unfixed gaze, and as a *coup de grâce*, it fouls itself, releasing a slimy, green fluid from its anal glands that smells somewhat like rotting flesh. Even when touched, the opossum keeps up the death display, and it may stay in that state for as long as an hour. By that time, the would-be predator would have lost interest and wandered off, making it safe for the opossum to rise once again, and hopefully go wash the foul-smelling slime off its butt.

Playing Hognose Snake

If the opossum is the star of the death-feigning performance, the American hognose snake could be the mammal's reptilian understudy. When it is under attack or feels threatened, the snake flops over onto its back and goes limp. Adding its own creative touch to the act, the snake lets its tongue droop out the side of its mouth and, taking a page out of the opossum's survival-strategy book, releases a foul-smelling liquid from its anus to make the would-be predator believe that the snake's flesh is decaying. However, the hognose is not quite as impressive an actor as its marsupial thespian counterpart. Should the predator flip it onto its belly, the snake quickly rolls over onto its back again, to continue—while at the same time effectively ruining—its performance.

Less Than Tasty

Feigning death to escape the real thing is not just the domain of mammals, reptiles and amphibians—some insects do it as well. The blister beetle is yet another creature that counts on its acting ability to keep it alive. However, this beetle must not have a lot of faith in its performance. As an added precaution, the beetle releases a foul-tasting chemical from its leg joints during its death portrayal, so that if the predator isn't dissuaded by the feigned death, it will be by the foul taste.

Going Down

Although they are common on lakes and ponds throughout much of Canada, the pied-billed grebe is not often seen. Perhaps this is because, when it feels threatened, this little grebe lowers itself in the water, rather like a submarine.

If it sees a predator lurking, the grebe stealthily drops out of sight by emptying its air sacs and squeezing out the air that is normally trapped between its feathers and its body. The grebe controls how high it sits in the pond or lake by altering the amount of water it expels from its plumage. It can choose to lower itself only slightly, sink until nothing but its head shows, or even swim along with only its eyes, beak and a bit of its tail remaining above the water's surface. No wonder this little grebe is so hard to see in its natural environment!

Ready for Takeoff!

Let me tell you, you don't want to stress out a vulture. If one of these carrion-eaters feels threatened, it will often throw up its stomach contents, making sure to place the putrid pile between itself and its would-be attacker. Not only does the smelly mass generally put the predator off (by killing its appetite with the stench, would be my guess—I'm not sure what kind of nastiness might be in the vomitous mass, but considering the birds' diet, it's probably even less pleasing the second time around), but by ridding itself of its last meal, the bird makes itself lighter in case a hasty takeoff is necessary.

Bloody Battle

In the landscape the horned lizard calls home, there is not a lot of cover for the creature to conceal itself in should a potential predator happen by. And many predators find this lizard tasty; cats, hawks, snakes and ground squirrels will all devour a horned lizard, given the chance. But this plucky lizard has a secret, and rather grotesque, defensive tactic—it squirts blood from the corners of its eyes. When it is under attack, the lizard increases the blood pressure in its sinuses until the sinus walls burst and blood comes shooting out from its eyes. The lizard usually aims the flow at the assailant's face, preferably the eyes or mouth, and can hit a target up to one metre away. The face full of blood doesn't hurt the would-be attacker, but it is usually shocking enough to make it change its mind and find a less-repulsive creature to snack on.

DID YOU KNOW?

Horned lizards have another neat defensive behaviour. If they choose not to shoot blood from their eyes, they can often drive off a predator by inflating their bodies, making themselves look much larger and more threatening (though, in this state, they have been described as looking like spiny balloons, which doesn't really sound threatening at all).

Safety in Numbers

Have you ever looked into the sky to see a cloud of small songbirds chasing, dive-bombing and generally harassing a much larger bird of prey? Although it seems as though the smaller birds must have a death wish, their mission is actually one of aggressive self-defence rather than suicide.

When a group of smaller birds mobs a larger bird, they are working on the assumption that there is safety in numbers. A bird flying alone is much more likely to be taken out by a predator than a bird that is part of a group. Although mobbed birds sometimes kill a few of the smaller birds harassing them, more often than not, the larger bird just gets fed up with the commotion and leaves the area. So when a small songbird spots a predacious bird lurking in its territory, it sounds the alarm, often attracting other birds from neighbouring territories to join the fight. Sometimes a number of different species respond to the call and pitch in. After all, no one wants a predator on their doorstep, and if a bird doesn't join in the group effort to drive an unwanted guest away, it may soon find itself battling the predator one-on-one should the intruder be driven from its current perch and into the smaller bird's territory.

THE BETTER TO CATCH YOU WITH

Spider Spit

Weak eyesight and a slow gait are probably not a great combination when you are a small, terrestrial predator, but those are the cards the spitting spider was dealt. And as with many things in nature, where there is weakness, something stronger evolves to fill the void. With its poor eyesight, the spitting spider depends on its sense of touch to find food. The spider ambles along, feeling its surroundings with its front legs, which are highly sensitive and longer than the other legs. When it locates a potential food item, it sneaks as close as possible without being seen, then lets the spit fly! Yup, that's right—it traps its prey by spitting on it. This arachnid's spit is a sort of venomous silk that becomes a gummy mess as it reacts with the air. When the gluey strands hit their target, they basically paste it to the ground. Then the spider can saunter over at its leisure to finish the job and deliver a killing, venomous bite. The spider can spit a distance of more than five times its body length. Without its special spitting capabilities, this relatively small spider wouldn't be able to catch many large prey species—it is too slow to run down its prey, too small to overpower many creatures, and its jaws cannot open very wide, so it couldn't even use its venom to dispatch an animal that wasn't already tied down. Spitting spiders occur worldwide, and there are more than 150 known species, though only one occurs in Canada.

Eating on the Wing

Little brown bats are a common sight in the Canadian evening sky, dipping and wheeling after mosquitoes, mayflies and other insects. When a bat locates a potential prey item, it pursues and catches the insect in flight, but rather than trapping the bug with its mouth, the bat actually catches it with its wingtip. It then moves the captured insect back to its tail membrane and curls the tail forward to shove the bug into its mouth. As it performs this feat, the bat is flying through the night sky at speeds of about 30 kilometres per hour.

Smooth Moves

I'm sure you've heard of a snake charmer, but how about a rabbit charmer? Well, such a thing exists…sort of. The short-tailed weasel, also known as the ermine or stoat, preys on many insects and small rodents, but it has also developed a taste for rabbits. Rabbits may be good eating, but they are not easy to catch. There is a reason why we

have the expression "quick like a bunny." Although the weasel is spry and can take down prey that is larger than itself, rabbits are a challenge. So the sly weasel has developed a strange and unique way of rabbit hunting. First, it tracks down a rabbit den, or warren. Next, it puts on a little dance, jumping, tumbling and spinning in a maniacal boogie. The rabbits seem to be more curious than afraid (and really, who wouldn't be?), so instead of running from the demented performer, they gather around for a closer look. As it spins, the weasel sneakily advances towards the rabbits until it is close enough to pounce on one and kill it. Then the show is over.

Blowing Bubbles

When you eat as much as a humpback whale does, hunting for food one fish at a time just doesn't make sense. The clever humpback has also reached this conclusion, which is why it has developed a strange way of catching fishing en masse—it blows bubbles. The whale rounds up a few of its humpback pals, and they work together, one herding a school of fish while the others swim in a spiral below the school. As they swim, the whales blow air from their blowholes, which comes out as bubbles. The bubbles form the shape of an upside-down tornado as they rise through the water column, surrounding the fish and effectively trapping them because the fish will not swim through the bubble walls. The whales then swim up the centre of the vortex and swallow the trapped fish by the mouthful.

DID YOU  KNOW?

The shape and colour of a humpback's tail flukes and dorsal fin are unique to each individual whale and can be used to determine its identity, much like fingerprints are used to identify people.

Wolves of the Sea

The much-maligned orca has been given a bum rap. These whales are not the bloodthirsty killers they have been made out to be, and there has never been a recorded case of orcas in their natural habitat attacking and killing a person. But that doesn't mean they aren't excellent hunters. These so-called killer whales are not actually whales at all; they are members of the dolphin family, and as such, they are some of the most intelligent creatures in the ocean. This keen intelligence is obvious in how they capture their prey.

Orcas are such efficient predators partly because they hunt cooperatively in packs and partly because they modify their hunting strategy according to the prey they are after. If they are after fish, they slap their fins against the water and call loudly so that the fish huddle together protectively; then the orcas snap their tails, and the resulting sound creates a pressure wave that stuns the fish long enough for the orcas to swim through the shoal and swallow them. Orcas have also been known to lunge onto beaches to grab young seals or sea lions, then use their front flippers to manoeuvre backwards until they are off the land and back in the water. Mother whales have even been witnessed teaching their young this strategy. They nudge the baby onto the shore, then grab it by the tail and help it work its way back into the ocean.

These clever predators have also been seen tipping seals off ice floes—the orcas dive under the ice floe in unison, making a huge wave that unbalances the ice floe and dumps the seal into the water, where the orcas are waiting.

And don't think that other whales are safe—a pack of orcas can even take down a full-grown blue whale. To kill a whale, the orcas work together to prevent it from surfacing to breathe. Before long, the unfortunate creature drowns and is then an easy (and substantial) meal for the pack. This strategy works well for most whales, but when orcas are hunting sperm whales, they have to modify their strategy a bit. When attacking a sperm whale, which can dive to depths far out of the orcas' reach, the pack keeps it at the surface, some individuals swimming behind the whale and some below it to keep it from diving, until they have completely tired it out. Then they go in for the kill. Researchers who study orcas are constantly surprised by the cleverness and adaptability of these creatures. Maybe instead of "killer whales" we should change their name to "cunning whales."

Pucker Power

The beluga's flexible lips do more than give this engaging creature a perpetual smirk. They are an important part of the whale's feeding strategy. Most of the foods the beluga eats, such as molluscs and crustaceans, are found on the seabed. The whale first locates its prey by using echolocation. Once it has homed in on a tasty morsel, the beluga purses its lips and sucks the prey into its mouth. Sometimes the soon-to-be-swallowed animal is a bit more troublesome to catch because it is stuck to the substrate or is hiding under stones or other debris on the seabed. In this situation, the whale doesn't really want to use its powers of suction because it would get a mouthful of grit as well as food. So the tricky

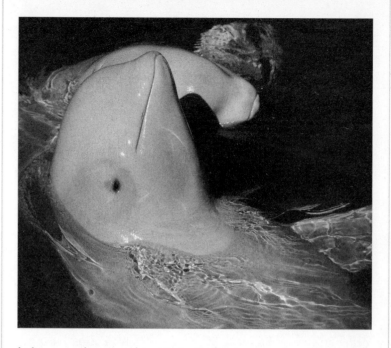

beluga employs another tactic. It draws water into its mouth, takes careful aim and squirts a strong jet at the creature, blasting it out of its hidey-hole.

A Sucker for Molluscs

I wonder if the walrus eyes the beluga whale's flexible lips with envy. Like the beluga, the walrus hunts its prey (mostly clams, but any mollusc or crustacean will do) on the sea floor. Also like the beluga, the walrus sometimes has to resort to squirting a stream of water to dislodge its prey from the substrate. However, the jet of water doesn't come from the animal's mouth—the walrus shoots it out of its nose. The blast of water unearths any creatures hiding in the sediment and sends them spinning into the water column, where the walrus can easily snatch them with its mouth. Once the walrus has a firm grip on the mollusc's slippery

shell, it sucks out the creature's soft insides by quickly pulling its tongue into the back of its mouth, creating a vacuum.
As you can imagine of an animal of its size, the walrus has a hearty appetite and can eat more than 5000 clams in one sitting.

Deadly Drool

Usually when one thinks of venomous creatures, one pictures the black widow spider or any of the world's many poisonous snakes, such as the cobra or the black mamba. I think it is pretty safe to say that, for most of us, a mammal does not come immediately to mind. However, the short-tailed shrew should be added to your list of known venomous creatures. It has poisonous saliva, so that with a bite, it can take down prey that would otherwise be too large for it to manage. The shrew's diet consists mainly of small worms, spiders, millipedes and insect larvae, but thanks to its toxic drool, it can overpower animals as large as a mouse. Toxins in the saliva damage the prey's red blood cells and affect its nervous system, paralyzing the animal. Depending how much toxin is chewed into the prey, it could kill the animal outright, or it could paralyze the animal without killing it, keeping it "fresh" until the shrew is ready to eat it.

DID YOU KNOW?

Unlike most of its kin, the short-tailed shrew also climbs trees, where it can snack on insect species that it would not find underground.

Leaping Egrets!

Cattle egrets are well known for their habit of following livestock or farm equipment, snacking on the insects that are flushed from the grass as the animals or machinery move through it. But what is a bird to do when there are no mammals or machines to follow? These egrets have developed a clever feeding strategy for just such an event—they play leapfrog. As the birds feed in a flock, the ones bringing up the rear fly over the backs of the ones in front, landing ahead of them and flushing the insects in the grass for the birds behind to grab. Once these insects have been dispatched, the birds in the back return the favour, flying over and landing ahead of the front line. And so it goes until every egret has eaten its fill.

Tongue Firmly in Cheek (Well, in Mouth)

A fly buzzes aimlessly through a swamp, no real destination in mind, just sort of flitting about. Suddenly a sticky, rubberband-like tongue snatches it out of the air and hauls it back to an open mouth, where it is swallowed whole. That's how a frog catches its prey, right? Not the tailed frog. In most frogs, the tongue is attached at the front of the mouth and is then folded backwards towards the throat, allowing the unattached end to shoot out and quickly snap back on demand. The tongue of the tailed frog, however, is firmly attached to the floor of the mouth, like a human tongue. Just as we can't shoot our tongues out of our mouths, neither can the tailed frog. Instead, this amphibian uses its body the way most frogs would use their tongue. It waits in ambush for an unsuspecting creature to wander past, then it lunges forward, keeping its feet rooted in one spot, and overshoots its prey, snatching it with its mouth on the way back as it recoils into a sitting position.

Drilling for Dinner

It can't be easy to open shelled prey when your entire body, except for a small beak, is made of soft, squishy matter. I suppose with its eight handy arms, the octopus could try to pull open the shell or hammer it against an underwater rock or something, but it would need a lot of force to battle the resistance of the water. Fortunately for the octopus, it has a better way. It "drills" through the shell. The octopus has special saliva that can soften the hard outer surface of a shell. As the saliva works its magic, the octopus uses its beak to scrape away the softened material until it has made a hole large enough through which it can secrete a special toxin that paralyzes the creature inside. The toxin also dissolves the tissue that holds the shell together, so after a short wait, the octopus can easily pull the shell open and snack on the innards.

Putting a Spin on It

Red-necked phalaropes are unusual little birds. Although they belong to the shorebird family, phalaropes turn up their beaks at the foraging style of their avian cousins. Rather than zipping along the sand searching for invertebrates, they take to the water. And rather than chasing after their prey, they make their prey come to them. Basically, these unique birds spin to dredge up a meal. The phalarope swims in a tight circle, spinning like a top (or as one researcher put it, "in a manner reminiscent of a slightly demented toy"), which creates a vortex that brings loads of tasty invertebrates to the water's surface, where the phalarope can pick them out of the water with its bill. When they spin, the birds do not stir up the sediment from the bottom of the water body; the birds spin so fast that they basically create their own upwellings, displacing surface water so quickly that the water underneath flows upwards to replace it without disturbing the sediment below. Also, every spin is completed a little to the side of the previous one so that the birds are constantly churning up new water, and therefore more prey.

YOU'RE GONNA EAT THAT?

Essence of Skunk

Any Canadian who has ventured out of the city and into the great outdoors is likely familiar with the smell of a skunk's spray. The stench is hard to miss and even harder to ignore. Now, it is bad enough having to smell the musk, but can you imagine having to taste it? That's gotta be unpleasant—but the great horned owl doesn't seem to mind. This large owl is one of the only predators that is courageous (or foolish) enough to take on a skunk. There's no telling how often the owl gets sprayed for its efforts, but it doesn't seem to faze the bird at all, because skunk is quite frequently on its menu. You'd think that when a skunk sprays, a little of the musk stench must permeate its flesh and fur, but who knows—maybe skunk musk tastes better than it smells. I can guarantee you, I'm not about to find out any time soon.

Better the Second Time Around

Consisting mainly of grasses, herbs and forbs, a rabbit's diet is not an easily digestible one. Much of the food passes through the digestive tract before all the nutrients have been extracted. Rather than eating huge quantities of vegetation to meet its nutritional requirements, a rabbit takes a rather distasteful shortcut—it eats its own poo. A rabbit produces two kinds of droppings: one that is hard and dry, and one that is soft and moist. The dry pellets are made up of waste material that the rabbit has no use for, but the soft droppings are rich in nutrients. By eating the moist pellets, the rabbit makes sure it gets the nutrients that its body missed the first time the food passed though, and it also saves itself the effort and energy required to forage for more food.

> **forb:** a herbaceous flowering plant other than a grass

Woodsmen Beware

Porcupines are not a camper's best friend. Because they have a taste for salt, and sharp teeth that are designed for chewing through tree bark, porcupines can be somewhat unwelcome on campsites, where they have a habit of gnawing on axe handles, shovels, canoe or rowboat oars—basically any wooden item that a sweaty-handed person might have handled. The rodent makes short work of the wood, extracting the salt residue left behind by the sweat.

Extreme Eating

When it comes to food choice, striped skunks apparently like to live a little on the wild side, choosing foods that many other animals avoid. For starters, they eat honeybees—the skunks break into the hives and scoop out the inhabitants, seemingly unaffected by the angry bees' stings. They also eat hairy caterpillars, the kind with the bristles that most animals avoid because the hairs irritate the digestive tract and can even be poisonous. Skunks have come up with a clever way of getting past this problem, though. Using their forepaws, they roll the caterpillars on the ground until all the hairs have been rubbed off. Also, being rubbed into the dirt seems to remove some of the toxins from the caterpillars' skin.

Got Milk?

Have you ever heard of a bird that feeds its young milk? It's true; there really is such a bird…well, sort of. For the first few weeks of their lives, nestling rock pigeons do not eat solid food. Instead, their parents (yup, both of them) produce a yellowish, cottage-cheese-like substance in their crops (a kind of stomach), which they regurgitate for the young birds. The substance, called crop milk, is actually the inner lining of the crop, which has been sloughed off. The resulting liquid is extremely high in both fat and protein, allowing the baby birds to grow quickly in their first few weeks. When it is time for the nestlings to switch from a liquid to a solid diet, the parents make the transition a little easier by feeding them food that has been softened in crop milk.

Spines, Schmines

The porcupine may or may not be a tasty creature, and most predators are not tempted to find out—they shy away from the spiny rodent, hoping to avoid getting a face full of quills. Not the fisher, though. This large member of the weasel family seems to have developed a taste for porcupine flesh, and it has perfected a method of killing its spiny prey that pretty much assures it (the fisher, not the porcupine, obviously) will come through the process unscathed.

Porcupines spend the majority of their time in trees, sleeping away the day or happily munching on bark, and pretty much the only time they are on the ground is when they are making their way from one tree to the next. If a fisher is lucky enough to catch a porcupine on the ground, its job is a little less laborious. The predator will circle the porcupine, just out of reach of that nasty prickly tail, and lunge

towards the rodent, snapping at its vulnerable face. Eventually, the porcupine gets too tired to keep circling, and the fisher moves in and flips it over, exposing its unprotected belly, and kills it. The fisher's strategy for treed porcupines is basically the same except that the pursuit takes place above the ground and the fisher has less room to manoeuvre, so the process is a bit more challenging. However, it can also be quicker if the fisher is lucky enough to unbalance the porcupine so that it falls from the branch and plummets to the ground.

Repulsive but Nourishing

As you may remember from an earlier chapter in this book, fulmars produce a smelly stomach oil that they spit at intruders who approach too close to the nest. What is really weird, though, is that this repulsive oil is not only used for self-defence but is also a vital food source for the chicks. The oil, which the parents feed to the young by regurgitation, provides the growing chick with nutrition that it cannot get otherwise in its diet. It also slows down the baby bird's digestion and acts as a food preservative to keep the chick going during the long spells that its parents are away from the nest, searching for their next meal.

Good Girl Gone Bad

Female black-tailed prairie dogs become somewhat nasty during the breeding season, killing and eating the young of other members of their coterie. A coterie in prairie dog culture is made up of closely related individuals, so basically the females are killing (and consuming) members of their extended family. Researchers suggest that the females choose to snack within their family circle because it is easier to get close to a baby within their own coterie than it is to try to sneak past the defences and protective mothers in a coterie to which they do not belong (and are therefore not welcome). What makes this cannibalistic behaviour even more odd is that black-tailed prairie dogs usually do not eat meat; for the rest of the year, the females' diet consists of grasses or other forms of vegetation.

Battle of the Titans

Why any creature would want to take on a giant squid when the oceans are teeming with much smaller, easier to capture prey is beyond me, but then I'm not a sperm whale. I guess it has its reasons. No one has ever seen the battle between these two giants of the sea, but the thought of it has captured the imagination of researchers and marine-life enthusiasts for ages. How does the whale do it? How does it take down the cephalopod behemoth? Giant squid are believed to be quick, agile and super-aggressive; scars on sperm whales' skin made by the suckers of giant squid prove that the squid is more than able to handle itself. However, squid carcasses retrieved from the stomach contents of whales also prove that they some-times emerge the victor. So what takes place in the ocean's murky depths? Until a battle is actually caught on tape, it will have to remain a mystery.

Move It or Lose It (Literally)!

Mormon crickets are not actually crickets—they are katy-dids. These insects are infamous for the hearty appetites they demonstrate when they are swarming. In fact, they get their name from an incident in Utah in which the insects swarmed through a Mormon settlement, devouring everything in their path. If not for the lowly gulls that swooped in and feasted on the insects, saving the day, the Mormon crickets would have wiped the entire settlement off the map.

These insects range from southern Canada to northern New Mexico. They generally feed on a variety of vegetation but have a weakness for crop plants such as wheat and barley. However, from time to time, they change their diet and move on to another food source—each other. Why a typically solitary vegetarian creature should suddenly begin to march en masse and develop a taste for the flesh of its own kind puzzled researchers for ages, but they have recently unearthed what triggers the change in this species. When Mormon crickets in an area become deficient in salt and protein, they start to swarm, seeking foods high in these nutrients, and any cricket that doesn't keep far enough in front of the one following it runs the risk of being eaten. After all, why waste time scouring the landscape for the nutrients you crave, especially when your usual herbivorous diet contains only trace amounts of these nutrients, when a good source is hopping just a few centimetres in all directions around you? During a swarm, this cricket has been known to eat the entire body of another cricket the same size as itself. Can you image eating a meal in one sitting that is the same size as you? According to researchers, this cricket is unique among animals that chew their food for being the only one that can accomplish such a feat.

Looking Good, Brother

Water is a scarce resource in the plains spadefoot toad's habitat. This toad's life cycle begins in water that pools in ditches or dips and depressions after it rains—hardly a stable environment for a developing tadpole. Because the pools are only fed by rainwater and do not stick around for long after the rain has stopped falling, the tadpoles must grow quickly. To do this, they need a lot of energy and nourishment, which most of the tadpoles will get by eating plankton or detritus. Notice I said "most." Occasionally, a tadpole hoovering up plankton will swallow another tadpole or a freshwater shrimp. Then everything changes; a whole new world of food possibilities opens up for the tadpole. Suddenly plankton and detritus don't seem so appealing, and instead, the mini-amphibian turns its palate exclusively to meatier prey, including fellow spadefoot tadpoles. The change isn't only in the cannibalistic tadpole's food preference—it also affects the animal's physiology. After the tadpole swallows the initial fleshy meal, the musculature in the tadpole's head changes, as does the structure of its gut, so even if it wanted to, it couldn't go back to its omnivorous diet.

Beware the Bigheads

If they knew what I know, all larval tiger salamanders would shun their big-headed brethren and hang around with the regular, small-headed kind. All tiger salamanders are born with standard-sized heads, and those that develop in areas where the population of larval tiger salamanders is low stay that way. But larval salamanders that live in crowded areas sometimes develop bigger heads and a taste for their own kind. The big-headed morphs turn up their snouts at the invertebrates that make up their typical diet and adopt a cannibalistic lifestyle instead. The (potentially) good news for larval tiger salamanders sharing the same habitat is that the cannibals recognize family ties when choosing their next meal. They tend to snack on individuals with the lowest levels of relatedness, taking non-relatives first and then preferring distant relatives over close relatives.

Don't Eat and Fly

With their pudgy bodies, spiky hairdos and black, bandit-like masks, cedar waxwings are a favourite with many bird enthusiasts in Canada. These endearing birds feed mainly on sugary fruit and seem to be especially fond of mountain-ash, at least in their Canadian habitat. They flock to our backyards in summer, trilling happily as they help themselves to the berries that have remained on the trees over winter. The fact that they are snacking on last year's berry crop doesn't seem to bother them in the least. Perhaps it should. As the ripe berries hang around, they begin to ferment, and the sugar they contain turns to alcohol. Waxwings feeding on fermented berries demonstrate definite signs of tipsiness. They have been seen flying erratically, crashing into walls, trees and the like, and even falling from their perches to lay passed-out on the ground. So next time you see a waxwing rambling through your lawn, don't assume it is injured—it could just as well be drunk.

Oh, Beehave!

Speaking of drunkenness, bees don't seem to be any better at holding their alcohol than waxwings are. They also feed on sugary foods that can ferment as they sit around—only the bees eat nectar and sap instead of berries. After ingesting its fill of fermented sap, an intoxicated bee weaves its unsteady way back to the hive, often ricocheting off trees or other solid surfaces as it flies along. When it reaches the hive, though, the tipsy bee has to pull itself together. Like a rebellious, party-going teenager returning home to face a watchful parent, the drunken bee must put on a convincing act of sobriety if it wants to be welcomed back into the hive. All hives have guard bees on duty at the entrance, ready to

drive away unwanted visitors, and these "bouncer bees" do not take kindly to their intoxicated brethren. If it cannot pass muster, the unruly bee is given the boot, sometimes for good. If it tries to sneak back into the hive and gets caught by the bouncers, it can suffer rather harsh consequences—the bouncers may bite off a few of its legs.

Tiny but Filling

It constantly amazes me that the largest creature in the world—in fact, the largest creature ever to have lived on earth, including all the dinosaurs—survives by eating one of the smallest. You would think that, to sustain such a large body, the blue whale would eat large prey items—maybe other whales or a walrus, or even colossal squid—but that is not the case. The blue whale's primary source of food is krill, with the odd fish or other small crustacean thrown in for good measure. On its polar feeding grounds, the blue whale eats about 3 tonnes of krill every day to maintain its bulk. Once the whale heads for its warmer breeding grounds, food is less plentiful, and the huge creature has to rely mostly on the large layer of blubber it built up over the summer.

DID YOU  KNOW?

With its throat grooves fully extended, holding a mouthful of water, a whale is more than twice its normal size.

Beware the Butcher Bird

When you look at a shrike, either a northern shrike or logger-head shrike, it looks relatively harmless. I mean, it is not that big, not like a hawk or an eagle, and it is actually quite cute, with its black mask and stout shape. But lurking beneath that attractive exterior beats the heart of a killer. In fact, this bird is so ruthless that it has been nicknamed the "butcher bird." So what has earned this innocuous-looking bird such a reputation? Well, it is practically fearless in its pursuit of prey, and, like an avian Vlad the Impaler, it sur-rounds itself with the hanging corpses of the creatures it has killed.

The shrike is not a picky eater. It hunts a variety of insects, amphibians, small mammals and songbirds. It often kills small birds, such as goldfinches and siskins, and can take larger species as well, such as American robins and some shorebirds. There have even been reports of shrikes killing rock pigeons and attacking sharp-tailed grouse and mallards! A mallard is three times the size of a shrike and is more than double the butcher bird's weight!

Once the shrike has taken down its prey, it impales the unfortunate creature on a sharp vine or a barbed-wire fence, or wedges it into a V-shaped fork on a branch. This rather ghoulish behaviour has some very practical perks. If the prey wasn't already dead, impaling it does the trick, which allows the shrike to kill prey that would otherwise be much too large for it to handle. Also, by hanging carcasses around its territory, the bird can store food for when prey is scarce and show off its hunting skills to wow potential mates. Finally, the shrike impales toxic prey, such as monarch butterflies and a few species of toad, and lets it sit for a few days before eating it, giving the toxins time to degrade.

DID YOU  **KNOW?**

Shrikes seem to have a fondness for prey with stingers and readily prey on bumblebees, honeybees, wasps and hornets. They do not, however, like the venom that these creatures have in their stingers. Before eating its stinging prey, the shrike will bite the insect's abdomen near the sting gland just hard enough to squeeze the venom into the stinger, then wipe the stinger off on surrounding vegetation.

Hide and Seek

When you live in a land that is covered with a snowy blanket for a good part of the year, it can be a bit of a pain to hide your winter cache in the ground. When winter comes and you are hungry, you have to work twice as hard, first digging through the layer of snow and then trying to delve far enough into the frozen ground to retrieve what you've hidden there. That must be why the clever gray jay, also know as a whiskeyjack, turns up its beak at ground caches. Instead, it hides its winter stash in trees.

This jay uses its sticky spit to attach food to clumps of conifer needles and strands of lichen, or behind flaps of tree bark. In this way, no matter how much snow falls, the bird still has easy access to its stores. A gray jay will have hundreds of caches made up of such foods as insects, scraps of meat, seeds and basically anything this opportunistic bird can find (no doubt, in the mountain parks, the caches consist largely of fries, pizza crusts and any other snacks these little robbers have thieved from the plates of tourists). The jay's saliva is edible and works as a preservative, ensuring that perishable items do not rot if they aren't needed until months after they were stored, so when the bird gets hungry, all it has to do is locate its stash, pull a morsel off the bark and eat it as is.

Clean Freak

Our favourite little bandit, the raccoon, has amazing manual dexterity, on par with that of many primates. In fact, the bones of a raccoon's paws are almost identical to those of an infant's hand. With its nimble little digits, the raccoon can peel fruit, throw stones and open pretty much anything it sets its mind to—clamshells, nuts, trash cans, padlocks… well, maybe not, but who knows; its abilities are pretty darned impressive. One of this mischievous creature's habits has given rise to the widely held belief that raccoons don't drink water and get all the moisture they need by washing their food before they eat it. Although the no-drinking-water business is a load of rubbish, it is easy to see where the myth originated. The raccoon prefers to live in wet or marshy areas, and a large part of its diet is made up of aquatic animals—frogs, clams, snails, fish and even turtles. While searching for these creatures, the raccoon dabbles in the water, and once prey is caught, holds it underwater and

manipulates it with its front paws. Scientists believe that the raccoon is actually "seeing" the prey with its paws, trying to get a better sense of what it is about to eat. However, to a casual observer, the raccoon looks very much as though it is washing its prey.

Why Hunt When You Can Steal?

No matter what your opinion is of magpies, you can't help but be impressed with their daring. These rabble-rousers form a sort of tag team around larger predators to steal their prey. Hawks, bald eagles, even foxes and coyotes have been known to lose their catch because of the antics of these birds. One magpie will hop towards the predator, let's say a coyote, and tug on its tail. When the coyote spins around to face its tormenter, another magpie hops stealthily up behind it and again pulls its tail. This process goes on until one of the birds can either snatch the food and make off with it, or until the predator gets fed up and abandons its catch.

Not Just Your Average Bully

Bullying behaviour in humans is usually seen as the domain of people who have perhaps a little more going for them in the muscular development department than the intellectual department. Not so with one of the bullies of the sky. Although it is more than capable of hunting for itself, the clever parasitic jaeger prefers to let other birds do the hard work instead, and it has perfected the art of mid-air piracy. This seabird intercepts and harasses other birds in the air as they return from a successful hunting foray and either steals the food directly from their beaks or snatches it from their feet. Atlantic puffins, kittiwakes, gulls, terns and even birds that outsize the jaeger inadvertently feed the bandit. And the jaeger doesn't just go after birds that it sees carrying prey—it will also harass a bird that it has just seen feeding until the harried bird throws up what it has eaten, then the jaeger swoops in and gulps down the vomitous mass. Yum.

Fungus Farmers

When you picture a farm, what comes to mind? A vast, open expanse of land with crops? Maybe a barn, or some cows or pigs? How about a salt marsh on the Atlantic Coast tended by snails? Well, the last one wouldn't have immediately sprung to mind for me, but it's a valid image. The marsh periwinkle, a type of marine snail, feeds on a species of fungus that grows on saltmarsh cordgrass. To ensure that it has a constant supply of this fungus, the snails chew holes in the cordgrass, which the fungus then colonizes. Once the fungus has moved in, the snails can manage its growth, chewing more holes in the cordgrass as needed to keep the fungus population healthy.

Making Hay

Even on the hottest days of summer, the pika's chosen habitat is relatively cool; in midwinter, it is bitterly, paw-numbingly cold. The pika lives in mountainous regions, on talus slopes around or just below treeline. Winters are long and the land-scape barren, but this little creature does not sleep the season away. It shuts itself into its den, happily munching away on its cached food as it waits for spring to arrive. To make sure it has enough food to last the winter, the pika must plan ahead, stockpiling as much as possible during summer and autumn.

But this little guy doesn't feed on seeds or grain, as do most mammals that hoard food; its diet consists mainly of fresh

green grasses, wildflowers or other types of vegetation.
As anyone who has looked into the refrigerator crisper only
to find a bit of slimy sludge that may have once been lettuce
or other forgotten greenery knows, fresh green vegetation
doesn't stay fresh or green for long—certainly not long enough
to be appetizing months after it was collected. So how does
the pika get around this little problem? Simple—it makes
hay. As summer begins to wind to a close, the pika starts
gathering some of its favourite greenery and laying it in the
sun to dry. Once the vegetation has dried out, the pika gathers
up its hay and moves it to the cache site, where it builds hay-
stacks. The feisty little creature isn't above a bit of looting,
either. It will readily reap the benefits of another pika's
labour, making off with its hay and adding it to its own
stash for the tough times ahead.

TOOL USE

A Tough Nut to Crack

It seems that some crows have a taste for nuts, but extracting the tasty nutmeat can be quite a challenge when you are a creature with no opposable thumbs. Crows are clever, though, so it shouldn't be surprising that they have found a solution to the problem. Some individuals have figured out that they can wedge a nut into a crevice or somewhere similar so that it cannot roll around. The bird then holds a rock in its beak and pounds on the nut's hard outer shell until it cracks. This strategy is a bit labour intensive, though, not to mention a bit hard on the neck muscles. So other crows that are perhaps a bit less motivated have come up with an easier method. An individual that chooses the second nut-cracking technique flies to a great height, grasping the nut in its feet, and drops it from on high, letting the combination of gravity and hard pavement take care of the rest. The nutshell essentially explodes on contact with the concrete, and the crow then flies down and picks the edible bits out of the smashed shell. This strategy is effective, but it still seems a little too much like hard work for some of these clever birds, so they've taken the idea and added their own little twist. Rather than taking to the skies with the nut, a crow will strut onto a road and drop it where a passing vehicle can drive over it, smashing the shell, then the bird swoops in to reap the benefits.

Dubious Home Decor

It may seem like an odd choice of home decor to us, but to a burrowing owl, dung is all the rage. This little owl lines the opening to its burrow with the droppings of larger animals such as cows or deer. The piles of stool catch the attention of dung beetles—one of the burrowing owl's favourite prey items. The beetles are lured close to the burrow, so that when it is feeling peckish, the owl can simply stroll up to the tunnel entrance and tuck into the beetles gathered there. Clever and practically effortless—the bird barely has to ruffle a feather. I've got to say, though, that dealing with all that poo seems an awfully high price to pay for a beetle snack.

It's Hammer Time

A good deal of the sea otter's diet comes in a hard, protective casing. Shellfish, molluscs, sea urchins, crabs…the otter loves them all, but before it can snack on the soft, squishy insides, it must first get past the hard outer shell. Not a problem—the otter knows just what to do. It swims down to the seabed, grabs a tasty-looking specimen, resurfaces and rolls onto its

back in the water. Then the otter lays the prey against its belly, takes careful aim and hammers the shell with a stone that it grasps between its front paws. After a few well-placed blows, the shell cracks, and it is dinnertime for the otter. Because one never knows when hunger may strike, the otter often carries a rock tucked into its armpit for just such a purpose.

DID YOU  KNOW?

A sleepy sea otter will anchor itself to a bed of kelp, wrapping the long, grass-like strands around its body before having a snooze so that it is not at the mercy of ocean currents as it sleeps.

Fishing With a Twist

Why go to your food when you can make your food come to you? This is surely a question that the green heron has taken to heart. Rather than using the slow, methodical stalking technique that many of its heron relatives have adopted, the green heron often hangs out on a perch over the water's surface. As it sits, it drops debris such as twigs or leaves into the water. Fish, mistaking the vegetation for bugs, swim closer to investigate, and once they are within the heron's reach, it snatches them up and eats them.

WONDERFULLY WEIRD BEHAVIOUR

Incoming!

Any creature in the vicinity of an American silver-spotted skipper caterpillar nest might be wise to keep one eye on the sky. Otherwise, it might be in for a nasty surprise. The caterpillar has a tendency to shoot its feces away from its nest, and the little critter can fire the poo pellets distances of more than 1.5 metres—more than 30 times its own body length! But the skipper is not just a neat and tidy creature trying to keep its immediate surroundings refuse-free; it is actually trying to confuse predators that are attracted by the feces. Certain species of wasp are drawn to the scent of the caterpillar's feces, called frass, so if the poop is near the caterpillar, the wasp will find and most likely eat the caterpillar. If the feces are far away, so too will be the predacious wasp. So how does the caterpillar do it? How does it turn its droppings into projectiles? Well, it pumps up the blood pressure under a special plate on its back, then puts the fecal pellet on the plate and just lets 'er rip. The pellet is ejected at a speed of 1.3 metres per second, rivalling the speed with which a cannonball is blasted from the barrel of a cannon.

Slip Slidin' Away

Who says animals don't have fun? Play is an integral part of a young mammal's development; wrestling, chasing and play fighting all teach the youngster valuable life skills. But river otters go one better than the usual games of tag or wrestling—they like sliding. These otters are known for zipping down snowy or muddy slopes on their bellies into ponds or rivers; as they slide, they tuck their front paws into their sides and draw their hind legs up against their bottoms. Now, you might be thinking that the otters are just finding a quick means of getting themselves down the slope and into the water. It's possible, but this doesn't explain why, once they've splashed into the water, they swim back to shore, run back up the slope and slide right back down again. River otters do not outgrow this behaviour, either. Even the adults toss themselves down slopes with abandon.

GROOMING AND SELF-MEDICATING

Squeaky Clean

To keep their plumage clean, birds such as crows, blue jays, wood thrushes and northern flickers perform a behaviour called anting. The birds take advantage of the formic acid on an ant's body to rid their feathers of mites and bacteria. Some birds squish the ants with their bills, then run the remains through their feathers. Others, such as crows, simply seek out an anthill, lower their wing tips to the mound, fluff their feathers and let the ants climb aboard and stroll about. Starlings are a bit more aggressive with their anting style—they catch an ant and shove it between their feathers, herding it so that the panicked critter runs madly about, secreting formic acid in self-defence, trying to find its way to freedom.

Birds that do not have easy access to ants still engage in the behaviour, but they find other arthropods, such as millipedes, to use as substitutes (these birds use the crush-and-smear method rather than letting the millipedes run amok on their bodies). If no formic-acid-bearing creatures are about, some birds will use less-appealing items instead (not that millipedes and ants are all that appealing, really)—objects such as cigarette butts and mothballs. Birds that prefer not to have ants or other unpleasant bugs running through their feathers take dust baths.

Bear Medicine

As the world around us becomes more and more industrialized, many people are cultivating an interest in things that are a little closer to nature. The popularity of traditional medicines, for example, is on the rise as people turn away from the synthetic compounds manufactured by pharmaceutical companies and gravitate towards alternative medicines with natural ingredients. Some of the plants used by indigenous cultures in their traditional remedies have been reintroduced into popular culture. But where did the Natives originally get their knowledge? In the case of osha root, the answer is from bears.

Osha, a member of the parsley family, grows throughout the Rocky Mountains at elevations of 2100 metres or higher. When they first emerge from hibernation, many bears head directly to the nearest osha plant and dig up a few roots to munch on. Natives believed that bears were using the plant to cleanse their digestive tracts after a prolonged period of inactivity. This theory has been given credence by the fact that bears also eat the plant when they are sick or injured, and by the fact that in traditional medicine, the plant was used for its antiviral and diuretic properties, as well as to stimulate the immune system. Bears also use the root to keep their fur clean and parasite free. They chew the roots of the plant into a pulp, then spit it onto their paws and run the squishy mess through their fur. Or, if that seems like too much work, the bear may just roll on the plant, covering its fur with the scent.

Vitamin Fix

A rabbit cleaning its ears is not just being hygienic—it is also setting the stage to ensure that it gets a vitamin boost. As a rabbit grooms it ears, it works a special oil into its fur. This oil contains a particular chemical that is broken down by sunlight to provide the rabbit with a good dose of vitamin D.

Keeping it Clean

Ants have a special gland, called the metapleural gland, that secretes an antibiotic fluid onto their bodies. The antibiotic seems to be released from the metapleural gland when the ant grooms itself, stimulating the gland, or when individuals in a colony groom each other. Then, as the ants move about the tunnels of their colony, the antibiotic is passed from their exoskeletons to the soil, wiping out any bacteria or fungus that may be trying to establish itself and keeping the colony disease free. However, the jury is still out on whether the ants purposefully spread the antibiotic (that is, by choosing when and where to groom) or whether the substance just oozes from their bodies and is passed to the ground with no conscious thought on the ants' part.

NOT GETTING ALONG

Nasty Neighbours

A female house wren will puncture the eggs of other birds nesting in the vicinity of her nest. She is not particular what species of eggs she punctures, either—any egg will do. Female wrens have also been accused of tossing other species' nestlings out of nearby nests. House wrens sometimes sabotage the nests of other birds by filling them with sticks, making them unusable and encouraging the other birds to move to friendlier territory.

On the upside, male house wrens have been seen feeding nestlings that were not their own, and in fact, were not even of the same species. It is not clear how common this behaviour is, but males have been observed feeding flicker and house sparrow babies.

Fight! Fight! Fight!

When they are not busy harassing neighbourhood cats or tricking hapless coyotes out of their hard-earned prey, mischievous black-billed magpies sometimes turn their attention to each other. Magpies live in relatively close quarters, for non-colonial birds. They can usually sort out their differences with vocal and visual signals, but when all else fails, the feathers fly. The two quarrelling birds jump through the air trying to kick each other. If one bird can get the advantage, it will grab the other bird by the feet so that it is unbalanced and falls on its back. Usually by this point, the kerfuffle has caught the attention of other nearby magpies, who come

swooping in to watch the fight. They surround the squab-
bling pair, squawking and calling, almost as though they are
egging the combatants on—much like a bunch of hyped-up
teenagers would when witnessing a street fight. If the action
gets a little dull because one of the fighting duo manages to
pin the other, one of the bystanders often hops into the ring
and pulls the dominant bird's tail, distracting it enough so
that the pinned bird can free itself and the battle can begin
all over again.

SOCIALIZING, ANIMAL-STYLE

No Click or Clack—Are We Under Attack?

Whenever a caribou takes a step, its foot makes a clicking sound. The click is caused by movement of the tendons—they spread as the caribou puts its weight on the foot and the foot flattens against the ground, then snap back into place as the caribou shifts its weight and lifts the foot to take another step. So every footfall has an accompanying click, sort of as though the animal were clacking its way across a wooden floor in hard-heeled shoes. (Can you imagine what it must sound like during the caribou migration, when thousands of animals trek in unison across the landscape? The mind boggles.) The clicking acts as a form of communication for the creatures. Even when their visibility is limited, say by fog or by the darkness of night, a herd can keep track of where its members are by the clicking of their feet. If they hear movement that isn't associated with a click, they know that an intruder, perhaps a predator, is in their midst.

Rude but Effective

Herring, which live in North Atlantic and Pacific waters, have an interesting means of communicating. They, to put it politely, emit air bubbles from their, um, anuses. That's right—herring communicate by farting. With this communication style, the message can be received in two ways—through sound and vision. First, the sound. A high-frequency noise accompanies the bubbles as they burst from the fish's butt; any herring in the surrounding area will get the message loud and clear. Second, the visual aspect. Nearby herring can see the stream of bubbles as they shoot out of the fish. I don't know what kind of message the herring are communicating as they pass gas, but scientists have determined that the fish are indeed communicating. And for their research efforts, the scientists who added this little gem to our library of scientific knowledge earned an IgNoble Award—an award handed out annually by Improbable Research, an organization dedicated to collecting information that "makes people laugh and think."

Whale Song

Anyone who has heard one of those nature sounds CDs you can get from the card shop in the mall is familiar with the haunting sound of a singing humpback whale. A whale will sing for up to 24 hours at a time, and it has a range of *at least* seven octaves (the most a human can reach is six). Both females and males produce an array of sounds, including what have been described as grunts, clicks, twitters and even moos, but only the males produce the songs that have made the whales famous. The songs travel great distances, up to 50 kilometres, and any boats that are in

the vicinity of a singing whale will feel the notes as vibrations of the boat's hull. Pretty impressive for an animal that has no vocal cords. It is not clear exactly how the whales are able to create such noises without vocal cords, but the most widely accepted theory is that the humpbacks push air through the different tubes of their respiration system. All males in a region sing the same song, but the songs differ between regions. The males sing during the breeding season, so the purpose of the songs may be to either attract a mate or mark out the boundary of an individual's territory.

Orca Vernacular

Orcas are social animals, living in pods that can range from two or three animals to as many as 40. To communicate with other members of the pod, the whales use a variety of calls that, surprisingly, demonstrate a significant degree of regional variation. In other words, whales from different pods speak basically the same language, but with a different dialect. The dialects are so distinct that it is obvious, even to a human eavesdropping on the conversation, which pod a calling whale belongs to.

Listening In

Speaking of eavesdropping, harbour seals can tell from a whale's "chatter" whether it is a fish eater or if it prefers to snack on mammals, such as, oh, let's say tasty seals. With this knowledge, the seal can then decide whether it is safe to remain in the water or if hightailing it towards land may be a better plan of action. But how can one species decipher the communications of another? Well, at least one species of dolphin makes it easy for the seals. Transient orcas, which feed mainly on marine mammals such as seals and even other whales, have a different communication style than resident orcas, which eat mainly fish. Resident orcas live in large groups and are in almost constant communication with each other. The transient orcas live in smaller groups, usually no more than four or five animals, and they do not vocalize as much as the resident populations because they need to sneak up on intelligent prey rather than prey that basically works on instinct. Little do they know, the harbour seal is on to them...

More than Words

Also known as the sea canary because of its vast array of vocalizations, the beluga is one of the chattiest whales, and its trills, clicks, moos and squeaks can be heard both above and below the water's surface. However, it does not rely on sound alone to get its message across—it also uses facial expressions. The beluga has flexible lips that it can shape into a smile or frown, and it can even purse its lips enough to whistle. It also snaps its jaws together to make a drumming sound, which is thought to be a form of warning. The beluga is also a champion at lobtailing, which is thought to be another means of communication.

> **lobtailing:** the action by a whale of lifting its tail out of the water, then slapping it against the water's surface

A Quiet Caution

I'm sure most of us have seen our fair share of ground squirrels standing at attention beside the entrance to their burrows, shrieking at the top of their little lungs to warn nearby ground squirrels that a predator is in the area before darting underground to safety. What might come as a surprise, though, is that these little creatures also have a "whisper" warning that they use in times of imminent danger. The whisper calls have a frequency of about 50 kilohertz (as opposed to the 8 kilohertz of the regular cries), which puts them in the ultrasound range and makes these squirrels one of the few mammals to use ultrasound as a warning. The whisper calls cannot travel very far and are likely used to warn immediate family members of nearby danger without alerting the potential predator or other unrelated ground squirrels that live farther away from the family unit.

Singing Lessons

You might think that a bird's ability to sing is instinctive—after all, your dog doesn't need to be taught to bark, and a pigeon doesn't learn how to coo—but that is not the case for grasshopper sparrows. The young male sparrows learn their songs by listening to adult males bellowing out their melodies, and not only adult grasshopper sparrow males. Young males seem to take a bit from all the calls they hear, regardless of the species, then improvise to create their own tune. So basically

it doesn't matter what species of bird the young sparrows hear, as long as they hear some. In lab experiments, male grasshopper sparrows that were raised in a soundproof box and never heard an adult sparrow's song could only produce weird buzzing noises.

Keep It Down, Please

Humans are not good neighbours. Not only do we contaminate our environment with chemicals and waste, we also pollute it with noise—and all this affects the wildlife that is trying to share our world. In Alberta, male ovenbirds have trouble finding mates because the noise from gas pipeline compressor stations drowns out their mating calls. And the ovenbird is not alone; many of Alberta's migratory songbirds are facing the same situation. Even worse, noise pollution isn't limited to Alberta's boreal forest. In cities throughout the country, traffic noise and the din from our day-to-day lives drowns out urban birds' calls and masks the sound of approaching predators, putting the birds at risk. In response, some songbird species, such as blackbirds, song sparrows and house finches, have changed their calls, making them louder or higher pitched so that they can be heard over the background noise.

DID YOU  KNOW?

Birds are not the only animals that are being affected by the noise we make. Researchers studying the behaviour of orcas off the West Coast of North America have discovered that their calls are becoming much longer, apparently in response to engine noise interfering with their communication.

Large Predator, Seated, on the Left

Being such a tiny bird, the black-capped chickadee has
a lot of bigger animal foes that would love to make it a snack.
It has, therefore, developed a warning system of alarm calls
to alert other members of its species that danger is near.
The interesting bit, though, is that the chickadee modifies
its alarm calls depending on the predator it has seen, giving
other chickadees in the surrounding area precise information about the level of threat they face. When it spots
a predator in flight, the chickadee gives a shrill *seet* call;
when the predator is stationary, the *chick-a-dee* call is
given. The number of *dee* notes that are uttered tell the
other birds just how dangerous the intruder is. More dangerous predators, especially those such as pygmy owls, which
are agile and known to feed primarily on birds, warrant
more syllables, and the elapsed time between the *chick* and
the *dee* notes is shorter than for predators that pose less of
a threat.

Bilingual Birds

The nuthatch is similar in size to the black-capped chickadee,
and both species share the same habitat, so it only makes sense
that the two species would fall victim to many of the same
predators. However, the chickadee has a slight edge over
the nuthatch because chickadees live in cohesive flocks,
whereas nuthatches live in pairs. More birds means more
eyes to keep watch for danger. But the nuthatch has figured out a means of capitalizing on the chickadee's warning system—it has learned to speak chickadee. Yup, that's
right. Not only does the nuthatch react to the chickadee's
alarm calls, but it also seems to understand the information
that is encoded in them. According to researchers, the
nuthatch demonstrates the same level of anxiety upon

hearing an alarm call as do the chickadees for whom the call was intended, suggesting that the nuthatch realizes the extent of threat the predator poses without actually seeing the intruder—which it could only do if it understood the information in the chickadee's call.

Drum Duel

A common sound in coniferous woodlands across Canada is the hollow thumping of the male ruffed grouse's drumming display. The grouse drums to make his presence known, both to potential mates and to other males, and as a way of marking his territory. For his display, the grouse will clamber onto a log or tree stump. Next, he braces his tail for balance, then spreads his wings, rotating them quickly back and forth. The motion causes a temporary vacuum that fills with a rush of air, creating a mini sonic boom. The display begins slowly, with only two or three wing beats, then gradually increases in speed, until the grouse's wing movements sound like a drum roll. At times, two nearby males will have a drumming duel; the birds display one at a time, each listening to the competitor's display, then doing his best to out-drum his rival. Males can drum anytime of the year, but they mostly do it in spring.

Phreaky Pheromones

Wouldn't it be nice if we didn't have to rely on speech to communicate with each other? Just think, no more searching for the right words to express yourself, and all those awkward silences could be eliminated. Ants don't know how good they have it. They communicate chemically, through secretions called pheromones, and an ant uses its antennae to smell the pheromone trail left by other members of its colony.

Ants use a number of different pheromones to express a variety of information, but one of the most widely studied by scientists is the food trail excretion. When a foraging ant finds a good food source, it lays a pheromone trail so that it can find its way back to the source from the colony, sort of like a reverse of Hansel and Gretel, only using chemicals instead of breadcrumbs. Its fellow ants also follow the trail, leaving chemical markers of their own, making the trail even stronger and easier to follow, which attracts more ants. Once the food source has been depleted or if the trail becomes impassable, the ants release a different pheromone that tells the other foragers not to bother wasting their time following the trail.

Doing the Nectar Dance

Like fuzzy, winged interpretive dancers, honeybees communicate with other members of their hives through the movement of their bodies. Unlike interpretive dancers, however, the other bees in the hive actually understand what the bee is trying to express. The dancing bee does a fancy little jig in the shape of a figure eight, called the waggle dance. The dance changes in length and intensity depending on the information the bee is trying to convey. Lively dancing means the bee has found a good food source; less energetic dancing means it had no luck and will have to keep searching. If the bee has found a viable food source, it twitches its abdomen, modifying how long it continues to twitch depending on how far away the source is. Different bee populations put their own spin on the waggle dance, sort of like a dialect.

Dying to Communicate

Even in death, a carpenter ant is able to communicate with its fellow ants. When it is crushed, the ant's body releases a pheromone that tells the other members of its colony that it is dead so they can come collect its body and carry it home. However, when many ants are crushed, the level of pheromone that is released sends a different message to the colony. Ants that sense the increased level of pheromone go a little mad, becoming extremely defensive and attacking any unfortunate creature that crosses their path.

DID YOU  KNOW?

When a worker carpenter ant is disturbed, it rocks its body back and forth quickly, slamming its front mandible and rear end against the side of the nest. Its body can make up to seven strikes within 50 milliseconds. This behaviour, known as drumming, is a form of carpenter ant communication.

LOOKING FOR LOVE

Porcupine Passion

Apparently, what seems as though it should be a deadly insult is actually a come-on in the world of porcupine courtship. When a male porcupine stumbles across a female in estrus, he first tries to impress her with his dance moves. The two porcupines show off their fancy footwork while circling each other, all the while making a number of different calls and groans. If the dance segment of the courtship goes well, the male kicks his wooing up a level—he sprays her with urine.

Now, if someone trying to impress me sprayed me with urine, he'd get my attention, all right, but not in a good way. Let's just say that he'd better be pretty darned fleet of foot, especially if I had a nifty tail full of very sharp spikes. But a little urine shower doesn't seem at all off-putting to the female porcupine, and more often than not, she and the male will mate once the courtship ritual is done. After they have finished mating, the female quickly drives the male away (maybe if he had presented her with a tasty morsel of food instead of peeing on her, she'd have let him stay around a little longer…).

> **estrus**: a period of sexual receptivity and fertility in female mammals

Lobster Love

Lobsters have bladders in their heads. Yup, it's true. Even better, the female lobster uses the contents of her head bladder to attract a mate. When breeding time rolls around, the female lobster goes on the prowl, searching for the male with the best (that is, the safest) burrow. Once she has found her man, the female lurks for days outside the mouth of his lair. He isn't always as eager for her company as she is for his, so she has to win him over, which she does by spraying her urine towards his burrow and letting the water carry it inside where he is holed up, tempting him with her pheromones. The urine shoots out of her head, from ducts just below her antennae. When he can no longer resist her, he races out of his burrow, claws a-snapping, and the soon-to-be couple engage in a little boxing match. Afterwards, he leads her into his lair, and the real fun begins. To get things started, the male strokes his mate, sort of like a gentle massage, to loosen her up so that she can shed her exoskeleton. Once the shell is off, the mating begins. The female will hang out in her partner's burrow until her new shell hardens—then she strolls out of his lair and goes about her business, single once more.

Duped!

An adult male red-sided garter snake should be a little more careful in his choice of mating partners. Often the lovely little miss that he's chosen is not quite what she seems. She is, in fact, a he. A young male garter snake can release female pheromones from its skin to fool older male snakes into thinking it is a female and mating with it. Once the older snake is spent and has slithered away, the younger male is free to mate with the still-available female. A young male snake cannot compete with the larger adults for a chance to mate, so his little deception is likely the only way he would get near a female that is ready to breed.

DID YOU KNOW?

Garter snakes overwinter in dens called hibernacula in such high numbers that they form the largest concentration of snakes worldwide—one den can contain up 25,000 snakes.

> **hibernaculum:** a shelter in which a hibernating animal, usually a snake or insect, spends the winter

Rainbow Stickleback?

During the spawning season, the male stickleback actually changes colour. His typically greyish-green belly becomes bright red. These little fish usually stick together, but when it is time to go a-courtin', the male sets off on his own, digging a home for himself in the seabed. He also digs a few pits that he fills with vegetation glued together with a sticky liquid—not spit or anything (relatively) dignified like that; this fluid comes from the stickleback's kidneys. Once the male has built his nest to his satisfaction, he changes colour again. This time his upper body turns pale blue, letting the surrounding females know that this is one little fish that has his act together and is ready for a mate. To get a female's attention, he does a little dance, zipping back and forth in front of her and rubbing her with his spines, hoping to woo her into his love nest.

Love Floats

Not only do the pharyngeal pouches of a male walrus make handy flotation devices while he is napping, they also help him attract a mate. The air sacs play a dual role in courtship. First, they are responsible, at least in part, for producing the bell-like sounds the male walrus makes when he serenades prospective mates. Also, their flotation properties ensure that the male rests high enough in the water so that his characteristic song isn't drowned out by mouthfuls of salt water as he sings.

Ungentlemanly Suitor

Mating for a male salamander isn't so much about courting a female as it is about tricking her. The male is, shall we say, "anatomically challenged" when it comes to fertilizing a female—he hasn't got the necessary equipment. Therefore, he must fertilize the female's eggs indirectly. To do this, the male drops his sperm packet on the ground, then tries to coax a female to squat on it. If he encounters an unco-operative female, the male salamander gets devious—he drugs her. While he is courting her, rubbing his head against her skin, the male releases a chemical from glands under his chin that make her compliant. Then he can lead her to his sperm packet and nudge her into the proper alignment to receive it. The scoundrel.

All's Fair...

There are certain rituals that must be observed in court-ship for it to be successful. In the case of the scorpion fly, it goes like this—the male gets the urge to mate, so he tracks down and kills a tasty snack to offer to a female. Next, the male advertises his successful hunting and status as the resident stud by excreting a certain chemical that acts as a signal. Finally, the female decodes the signal and rushes to his side. Simple. However, there are always a few ne'er-do-wells that don't want to toe the line. In this case, lazy male scorpion flies that can't be bothered to catch their own prey also detect the successful hunter's signal and hover around him, waiting to intercept the females as they respond to the hunter's signal.

The Gift that Keeps on Giving

The male ornate moth, gentleman that he is, provides his chosen partner with a sort of life insurance when they mate. During his larval stage, the male moth feeds on rattlebox plants, which contain alkaloids; when absorbed into his body, the alkaloids make the moth distasteful to spiders that would otherwise love to snack on him. When the male moth finds a mate, he passes the alkaloids to his partner in his sperm package. But not all male ornate moths are created equal, and the female is quite choosy when it comes to deciding with whom to mate. During courtship, the male moth releases chemicals from two brushes on his abdomen, and these chemicals tell the female how much of the alkaloids he has to offer. Males with low alkaloid levels are given their walking papers; males with more impressive levels receive the go-ahead to mate. In this way, the female can be certain that if she gets caught in a spider's web in the future, the arachnid will not waste its time with her—it will simply cut her loose.

Preying, er, Praying Mantis

There has long been an understanding that the male praying mantis accepts its status as a post-coital snack when it chooses to mate, but this may not be the case (and really, does this come as much of a surprise to anyone?). Apparently, a female that is well fed before she mates is less likely to snack on her partner after the deed is done. Knowing this, the male alters his behaviour according to how hungry he perceives his mate to be. If she seems particularly peckish, he approaches cautiously from the side instead of head on, so she has less of an opportunity to grab him with her front (prey-catching) legs. He also tries a little harder to woo her, kicking the

courting behaviour into high gear. Perhaps to really increase his chances of survival, he should think about greeting her with a little snack in hand.

Less-than-Gentle Loving

Another less-than-ideal mating partner, in my mind, is the Dana octopus squid, only this time it is the female that gets the raw deal. This octopus is found throughout the temperate waters of the North Atlantic Ocean. During mating, the male approaches a willing female and slices her with his beak and claws. He makes a series of five-centimetre-deep gashes in her flesh and inserts his spermatophores (sperm packets). Then, if he has any sense, he beats a hasty retreat before she recovers from his rough treatment. I don't know what kind of mood she would be in after such brutish handling, but I'm thinking it can't be good.

Together Forever

I'm sure many a wife has thought that her husband is a pain in the backside, but never has this been truer than in the case of Ogcocephalidae anglerfish. There are 160 species of anglerfish in the Ogcocephalidae family, and they are distributed throughout the oceans of the world. These fish are not bottom-dwellers, but they do live in some of the earth's deepest waters, staying at least 300 metres below the water's surface. It's not easy to find a mate in the deep recesses of the ocean, so when the male anglerfish is fortunate enough to stumble across a female, he doesn't let her go. Literally. He sinks his teeth into her flesh, and because he has hooked denticles, he can really hold on. When he bites her, the male releases an enzyme that dissolves her skin around the wound as well as the skin of his mouth, and before you know it, male and female have become one. Like a parasite, he lives off her blood. Eventually, his circulation system meshes with hers, and his internal organs and eyes atrophy so that he is essentially nothing but a reproductive sac attached to her, like a pimple on her backside. By the time a male's degeneration is complete, a full-grown female can be about 500,000 times heavier than he is!

denticles: tooth-like projections that create an abrasive surface

REPRODUCTION

Breaking and Exiting?

Ever wonder why a gull or other predacious bird with a hankering for eggs can hammer its strong bill futilely against an egg's outer shell, giving up in frustration when it can't break through, yet a baby bird that doesn't even have the strength to hold up its head can bust its way out of the same shell with relative ease? It all comes down to the eggshell's design. The cells of an eggshell are aligned in such a way that the egg is easier to break out of than it is to break into. The calcite crystals that the shell is composed of are columnar and narrower on the inside, so that they fit together much like an arch. Because a chick breaking out of a shell is pushing from the inside, it forces the shell up and out—much like removing the keystone from an arch—weakening the structure. However, in attacking the shell from the outside, a predator is pushing down on the arch, which pushes the crystals together, reinforcing the strength of the shell.

Hang in There

How do you give birth when you are a creature that spends most of your non-flying time hanging upside down by your feet? Well, the little brown bat has a rather unique approach. She rights herself so that she is gripping the wall of the bat nursery with her wing claws and her feet are pointing towards the ground. Then, with gravity on her side, she waits for the baby to emerge and curls her tail membrane forward to catch it before it falls to the ground. This bat species usually has only one baby at a time, but twins do occur on occasion. Must be tricky to catch the second one.

No Time to Waste

The plains spadefoot toad lives in arid environments, but being an amphibian, it needs water during the early stages of its life. So what is a toad to do? Well, the adult male spadefoot waits until the first heavy spring rain, then emerges from its underground lair and hops madly to the nearest pool of standing water—a puddle or a ditch or whatever is close at hand. Once he has staked out his space alongside his watering hole, he begins his chorus to attract a mate. After the female lays her eggs, they float about in the water until they become attached to emergent vegetation, rocks or other objects in the water. Then the race is on! The eggs must hatch and tadpoles must grow and complete their metamorphosis into fully formed toads before the water evaporates and their puddle dries up. This entire process can take as little as two weeks, the fastest known cycle of any frog or toad.

What's the Hurry?

What do wolverines, bears, river otters, walruses and little brown bats have in common (other than the fact that they are all mammals—that's too easy). They all have a reproductive strategy called delayed implantation, meaning that the female's fertilized egg doesn't immediately implant onto the uterine wall, so the embryo does not begin to develop right away. Animals that use this strategy do so because there is not enough time between when the animals mate and the end of the female's gestation period for the young to be born into favourable conditions, such as mild weather or times with an abundant food supply. Delayed implantation allows the couple to mate and the young to be born when conditions are at their best, regardless of how long the female's pregnancy lasts.

DID YOU KNOW?

If a female bear is not in good enough physical condition to reproduce when she goes into hibernation, her fertilized eggs will be reabsorbed into her body.

Monogamy, One: Polygamy, Zero

Who says fidelity doesn't pay? For female bees (*Bombus* spp.), it can mean the difference between a long life and a much shorter one. Female bees that have been inseminated by more than one male die younger than those that have mated with only one male. It is not clear why this should be so, but the polygamous females seem unable to survive hibernation. It may be that they use up too much of their energy reserves while mating with numerous partners, or there may be chemicals in the sperm that are designed to attack the sperm of rival males but attack the female's system as well, weakening her.

Till Death Do Us Part

Now to focus on the males, lest you get the impression that they have it any better than the females. Male honeybees pay a high price for a little loving. During copulation, the male actually loses his genitalia. In a process known as autotomy, the drone's "equipment" separates from his body, and thus unmanned, he dies soon afterwards. The bit he left behind forms a "mating plug" in the female, blocking other suitors from mating with her unless they are able to remove it.

Fertilization from a Distance

The flour beetle has a neat reproductive trick—he can fertilize a female that he has never mated with. In the world of flour beetles, many males mate with a single female, and the male's body has adapted accordingly. To ensure that his are the genes that fertilize the female, the male flour beetle has spines on his genitalia that are designed to scrape away the sperm left behind by males that have been there before him, so to speak. Some of the scraped-away sperm sticks to the current amorous male's "equipment," and if he wanders off and mates with another female, he inseminates her with the scraped-away sperm as well as his own. For the rival's sperm to be viable, the subsequent mating must happen soon after the initial one, which doesn't seem to be an issue for these lusty beetles.

That's Favouritism!

All eggs are not created equal, at least not in the world of the cattle egret. First-laid eggs receive a higher dose of the male hormone androgen than the second-laid egg, and the second receives more than the last egg that is laid. The extra androgen makes the chick more aggressive so that it has an edge over its nestmates at feeding time. The chick that squawks the loudest, stretches its beak up the highest and bullies its siblings to the side is the one that gets the most food. Parents aren't concerned with the equal distribution of food they bring to the nest; they stuff the morsels into whichever open mouth they see first. So basically, the last-born chick doesn't really stand a chance. Not only is it bound to be smaller than chicks that have hatched earlier (and have therefore been receiving food for longer), it is actually genetically predisposed to be at a disadvantage.

In for the Long Haul

For a female lobster, pregnancy is a long, drawn-out affair. Up to 20 months can pass from the time she mates until the first mini-crustacean emerges from its egg. Don't worry, it's not as bad as it sounds. The female lobster doesn't necessarily fertilize her eggs as soon as she mates. Her body has a special receptacle where she can store the male's sperm for months, until she is ready to reproduce. When the appropriate time has come, the lobster lays on her back and cups her tail to release the eggs from her ovaries. The eggs are fertilized as they pass through the sperm receptacle on their way out of her body, at which time they get stuck in a gluey substance under her tail. The lobster mom will carry her eggs (sometimes as many as 20,000!) for nine to eleven months, fanning them with her swimmerets to make sure they get enough oxygen and to clean off any debris that might stick to them as they are developing. Finally, when the eggs are ready to hatch, the female lifts her tail and lets the current sweep them away from her body. The long event has come to an end at last.

> **swimmerets:** small appendages on the abdomen of many crustaceans that are used primarily for locomotion and, in females, for carrying eggs

No Brotherly (or Sisterly) Love

After mating, a female pronghorn may have four or more eggs implant into the wall of her uterus, but at the end of her long pregnancy, she will only give birth to two lambs. While they are developing in the uterus, the two fetuses closest to the top develop long spikes that hang down and penetrate the fetuses below—basically the fetuses stab and kill their womb-mates. How's that for sibling rivalry?

BRINGING UP BABY

Love Him and Leave Him

The female red-necked phalarope is an inspiration for women's rights advocates worldwide. This little bird has turned the whole "male meets female, male mates with female, male abandons female and goes off with other males, leaving the burden of responsibility on female" scenario on its head. She is the one who chooses and competes for her mate, and after she's had her way with him and has laid her eggs, she wanders off in search of another male that strikes her fancy. Meanwhile, her jilted mate is left behind to perform the domestic duties of incubating the eggs and tending to the young. As an outward representation of their role reversal, the male has the drab plumage, while the female is markedly more colourful.

Bad Parenting

Brown-headed cowbirds will not be nominated for "parents of the year" anytime soon. They can't be bothered to raise their own young, so they deceive some other hapless bird (quite a few other birds, actually) into doing it for them. First, the female cowbird lurks in the shrubbery or treetops, watching for activity from other birds in the area, particularly nest-building activity or birds that come and go frequently from the same location. Sometimes, when she gets a little impatient, the female takes a more proactive role, flapping around noisily in an effort to flush nesting birds. Once she has located a poor, unsuspecting nesting bird, she waits until it is out of the area and deposits her egg into its nest. She is then free to hang out with her bovine pals, snacking on the bugs the large beasts stir up. A female may lay up to 40 eggs

in one breeding season, and she will drop her eggs in the nests of many different species—more than 200 species have been recorded, including the ovenbird, the song sparrow and especially the yellow warbler.

A few clever bird species, however, are onto the cowbird and have learned to recognize when the larger bird has dropped an egg or two in their nests. The American robin will eject the foreigners, casting the offending egg out over the edge of the nest to smash on the ground below. The yellow warbler has a different strategy—it just abandons its nest entirely and rebuilds, often on top of the original one. If a bird doesn't realize that the sanctity of its nest has been compromised—and many don't—their own nestlings usually have a poor chance of survival. The bigger cowbird baby often outcompetes its nestmates for food so that they starve, and it may even bump them out of the nest. However, it is sometimes the cowbird fledgling that suffers. If the cowbird has chosen the nest of a vegetarian bird, such as a house finch, the cowbird youngster usually starves to death, regardless of how much food its adoptive parents provide.

Motherly Love

As an example of motherly dedication taken a little far, I offer the giant North Pacific octopus. The second largest known octopus in the world, this creature can grow to 9 metres in length (though 7 metres is more usual) and weigh more than 45 kilograms. The female is known for her fecundity and can lay between 60,000 and 100,000 eggs. After her eggs have been laid, she dedicates all her energy to protecting them, even foregoing food. By the time the baby octopuses emerge, the mother has wasted away to such an extent that the damage cannot be reversed, and she quietly slips away into the great beyond.

Fancy Footwork

During the nesting season, most female birds develop a brood patch so they can incubate their eggs. This patch is a feather-less area of skin in which the blood vessels are near the skin's surface, making heat transfer from the parent's body to the eggs possible. The feathers may automatically drop from the bird's body to expose the patch, or the bird may pluck the appropriate area on its belly.

In species where both parents share incubation duties, the male and female each develop the patch. In the case of the northern gannet, neither bird does. With no brood patch, the gannet has had to be a bit creative in its incubation style. To keep its egg at the right temperature, the bird stands on it. Heat is then transferred through it's big, webbed feet. Good thing the gannet lays only one egg at a time.

A Whole Lotta Milk

A blue whale mother produces about 200 litres of milk per day (almost enough to fill an average-sized bathtub), making her already substantial weight that much more impressive. Like all babies, her calf is basically an eating machine; however, with such large offspring comes an equally large appetite, and a baby blue will feed up to 50 times a day. Such intensive feeding ensures that the young whale grows quickly—it can pack on up to 90 kilograms per day!

No Rest for the Devoted

All new mothers have to sacrifice sleep to tend to their babies' needs, but an orca mother takes this to the extreme. She goes without sleep for an entire month after her calf is born. A mother with a newborn calf must be diligent; she has no time for such luxuries as sleep. Many predators would welcome the opportunity to snack on a newborn calf, so the mother must keep them at bay. She must also help her baby to the surface so that it can breathe and has to keep it warm until the (relatively) little one develops a thick enough layer of blubber to protect it from the cold ocean waters.

It would be physically impossible for a human mother to match the orca's extended period of sleeplessness. Sleep is a critical part of our lives; our bodies repair themselves as they rest. Long-term sleep deprivation can impair a person's vision and the functioning of the immune system as well as cause depression, hallucinations and paranoia. The female orca, however, doesn't seem to suffer any consequences— except maybe a deeply ingrained irritableness, but I'm just speculating. All I'm saying is, if I was responsible for keeping my mother up for a month straight, I'd be on my best behaviour until she got some rest.

DID YOU KNOW?

Those of you who think missing a little sleep is no big deal, be warned. Going for just 24 hours without sleep can be as incapacitating as being legally drunk.

Filling the Void…Somehow

There seems to be a bit of uncertainty regarding how a baby whale feeds. We know that the mother whale produces milk with which to feed her youngster, but we don't really understand how exactly the calf accesses the milk. If they were to give the matter any thought, I'm sure most people would just assume that the whale baby nurses like any other mammal baby does. Not so, say researchers. A baby whale's lips are not shaped in such a way that they would fit tightly enough around a nipple to allow for the suction necessary to draw milk into the baby's mouth. Another theory has it that the mother squirts the milk from her teat directly into her offspring's mouth, eliminating the need for suckling. The most outlandish theory that has been proposed to date suggests that the mother shoots her milk into the calf's blowhole, and the baby then swims to the water's surface to breathe and swallow. Hmmm…into the blowhole? Really? Clearly a little more research is needed to clear up this mystery.

Motherhood Transcends All Barriers

I'm not really sure if this counts as a weird Canadian animal, but it is one of my favourite weird wildlife stories, and it happened in Canada, so that's good enough for me and I'm including it anyway. For those of you who disagree that it has a place in this book, you can consider the duck weird for approaching a human in search of help (and for singling out a police officer, to boot) or the police officer weird for bothering to respond to the gentle persuasion of a duck.

On a lovely July morning in 2001, Ray Petersen, a Vancouver police officer, was strolling along Granville Street when a female mallard waddled up to him and grabbed his pant leg. Once she had his attention, she started circling him, quacking all the while. He pushed her away, but the stubborn duck could not be dissuaded so easily. Keeping an eye on him to be sure he was still watching, she padded over to a sewer grate about 20 metres away and lay down on it. Ray didn't take the hint. When he began to move away, she started the whole process over again—waddling, quacking and pant-leg tugging. This time when she headed back towards the grate, Petersen was right behind her. Peeping into the sewer, he saw eight baby ducks stranded in the water below. With a little help from some fellow officers and a tow truck (to move the grate), Petersen was able to scoop the babies to safety under the watchful eye of their mother. Once the family was reunited, the mother led her babies to the relative safety of some nearby water.

I LIKE TO MOVE IT, MOVE IT

It's a Bird, It's a Plane...

The octopus spends much of its time "walking" over the ocean's hard underwater surfaces, using its many arms as though they were legs, but this form of transportation isn't exactly a speedy way of getting around. Should the need arise for an octopus to put a little velocity in its movement, the creature gives up its creeping ways and uses a much more interesting mode of transportation—jet propulsion. The octopus draws water across its gills and into its mantle, then it seals off all bodily openings except the "siphon," a narrow, funnel-like structure attached to its mantle. To move forward, the octopus contracts the muscles in the mantle and forces the water out through the siphon. To steer, the octopus just has to change the direction in which the siphon is pointing. Using jet propulsion, the cephalopod can travel up to about 40 kilometres, but the burst of speed is short-lived. Once the water in the mantle runs out, so does the propulsion. Because it can't maintain this type of movement for long, the octopus generally only relies on it to snatch a nearby prey item or to quickly head for cover.

The octopus' jet propulsion also comes in handy, though, when trying to escape a predator; if the octopus can't outrun—well, outswim, really—its pursuer, it races to the ocean's surface and uses jet propulsion to shoot itself clean out of the water, leaving its baffled attacker in the water below wondering where the heck it went. An airborne octopus? Now that's something I'd like to see...

Swimming to Its Own Beat

As if it wasn't weird enough to look at with the giant horn growing out of its upper lip, the narwhal also has a strange way of swimming—upside down. This whale has a habit of swimming belly up, so that if you were to see it from the air or the water's surface, you would see the pale belly instead of the darker back. No one is sure why the narwhal swims this way. Also baffling to researchers is the whale's tendency to pause while swimming upside down and just hang out for a few minutes, floating in the water column. This behaviour gave rise to the whale's common name, narwhal, which comes from an Old Norse word meaning "corpse whale."

Stampeding Sea Cucumbers

All right, here is a mental image that you will not be able to get out of your head for a while. Galloping sea cucumbers—herds of them. It's true. The creatures that look more like produce than animals sometimes travel across the sea floor in large herds, rather like horses of the sea, grazing on microscopic organisms as they go. Unlike horses, however, sea cucumbers

can choose to float rather than gallop. These unique creatures are constructed in such a way that they can control
their own buoyancy, so if they get tuckered out inching
along the sea floor, they can always float into the water
column and let it carry them to their next destination.

Going for a Stroll

Anglerfish are just weird. They look weird. They act weird.
Even the way they move about in the water is weird. Unlike
most fish, which are happy to swim about in the water,
bottom-dwelling anglerfish turn up their strange-looking
barbels at such a pedestrian way of getting around. Instead,
they prefer to walk. These fish have specially modified
pectoral fins that basically take the place of feet, allowing
the fish to walk along the bottom of the ocean. Some
species also have modified pelvic fins to make walking
easier. You'd think a walking fish would be easy to spot,
but these guys make it tricky, hiding in sand or seaweed
when they feel threatened.

Falling with Style

I suppose, technically, the flying squirrel should be called
the gliding squirrel, but that doesn't have nearly the same
ring to it, does it? No air of romance. Despite its common
name, the flying squirrel doesn't actually fly—you could
think of its mode of transportation as more of a controlled
fall than actual flight. When this squirrel wants to get
from one tree to the next in a hurry, it basically throws
itself from a high branch and spreads out its limbs, letting
the air catch the flaps of skin that are attached to them.
As the air pushes against the skin flaps, they billow out and
slow the squirrel's fall, rather like a parachute. This squirrel

is quite manoeuvrable in the air. To change direction, it lowers the forelimb slightly on the side of its body that faces the direction in which it wants to turn, then lets the wind do the rest. As the wind catches the skin fold, the squirrel leans its body into the turn and uses its tail as a rudder, finessing its movements. It can travel about 9 metres in a typical glide, but glides as long as 40 metres have been recorded. Youngsters master the skill of gliding by the time they are about eight weeks old.

There are two flying squirrel species in Canada—northern and southern flying squirrels—and the northern species can grow to be 30 centimetres long! You'd certainly notice *that* if it came flying down at you (which, of course, it wouldn't—flying squirrels are nocturnal and are very timid, avoiding contact with people).

Moving Around Under the Ground

The northern pocket gopher has many weird adaptations that help it move about quickly and efficiently in the underground tunnel system it calls home. First, the creature has loose-fitting skin so it can easily manoeuvre in tight spaces. Second, the pocket gopher has special hairs in its coat that attach directly to nerves, much like a cat's whiskers. However, instead of being confined to the face, as they are with a cat, these special hairs cover the pocket gopher's entire body. And last, this animal's tail is bald at the end and is also highly sensitive to touch. The pocket gopher uses its tail to feel its way through its underground lair, and because the tail is so sensitive, the animal can run backwards through tunnels (using its tail as a guide) just as quickly as it can when going forward. This is clearly not an animal that relies heavily on its eyesight.

DID YOU  KNOW?

The same loose skin that helps the northern pocket gopher move about easily underground also gives it an edge when under attack. Because the skin is so loose, an aggressor that bites the pocket gopher often gets nothing more than a mouthful of fur and skin, missing the more delicate tissue underneath and allowing the pocket gopher to escape serious injury. Also, when grabbed by the back of the neck—say, for example, by a predator intent on eating it—the pocket gopher is not incapacitated as most animals would be; it can still turn around enough to defend itself and sink its teeth into whatever creature is holding onto it.

Who Needs Ice to Skate?

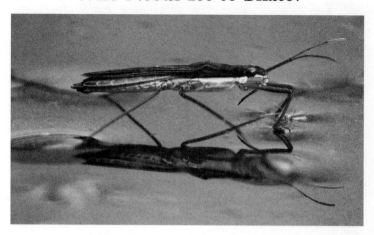

Surface tension is the water strider's best friend. Without it, the creature would not be able to skate across the surface of still water. The water strider has special hairs on its legs that trap air and repel water. It uses its two long, middle legs like oars to propel itself forward, its long hind legs to steer and brake, and its two short, middle legs to detect its prey.

Without surface tension, this critter would also go hungry. The two middle legs are kept in constant contact with the water's surface so they can feel when another creature breaks the surface tension in the surrounding water. As soon as it feels the tension being disrupted, the water strider races over and eats whatever caused the disturbance.

Slow but Unstoppable
The snail cannot be rushed. This slow but determined invertebrate will eventually get where it wants to go, despite any obstacles in its way—it can climb up fences, walls or whatever happens to be in its path and can even cross ceilings

while moving upside down. The snail moves about on its giant foot, which has a gland near the front that produces and secretes mucus to lubricate the snail's way. The mucus is unusual in that it is a non-Newtonian fluid, meaning that its viscosity changes, becoming thicker when it is subjected to stress. This change in the mucus' properties is what allows the snail to move about upside down on the underside of surfaces. To move forward, the snail contracts the muscles in its huge foot, which are aligned in such a way that they ripple in succession across the bottom surface of the foot, rather like waves in the ocean rolling to shore, carrying the snail forward like a really slow surfer.

On the Move

When they cannot get enough salt or protein from the food resources available in their habitat, Mormon crickets band together and relocate. This behaviour, which we laypeople generally call swarming but researchers call a forced march, allows the crickets to move in relative safety to better surroundings with food that will meet their dietary needs. By moving en masse, the crickets are less vulnerable to being attacked by predators (though this doesn't keep them safe from each other; see the Weird Foods section). However, unlike swarming locusts, Mormon crickets do not consume everything in sight, leaving behind a barren landscape. These katydids are much more selective in their food choices, and the destruction they leave in their wake has been likened to that of a tornado (in pattern, not intensity). Parts of the landscape will be left untouched, while areas that contained foods the katydids were seeking will be decimated. While swarming, these insects can cover more than 1.5 kilometres per day—not bad for creatures whose sole means of locomotion is a crawl. During bad outbreaks, the density of these insects has been as high as 100 katydids per square metre, and swarms have been recorded that were 10 to 50,000 strong.

DID YOU KNOW?

An American water shrew spends most of its life eating. It can actually starve to death if it doesn't eat for three hours. These shrews generally have a snack every 10 or so minutes and spend pretty much all of their waking time searching for food.

Move Over, Prophet Who Shall Remain Nameless

Mammals can't walk on water, right? Bugs? Sure. Reptiles and amphibians? Well, maybe a few. But mammals? They'd sink like large, fleshy stones, right? Ha! Guess again! Nothing can stand between a hungry American water shrew and a potential meal. If the little critter has a hankering for fish, it can swim or run underwater along the bottom of the stream to hunt. If aquatic insects are on the menu and the shrew isn't in the mood for a swim, it can simply run across the water's surface to snatch a bug. That's right, this little creature can walk—well, actually, run—on water. Stiff hairs on the shrew's feet trap air bubbles under each foot, keeping the animal afloat as it moves across the surface of a small pond or pool.

MIGRATION

Long-Distance Champion

The Arctic tern has earned the title of long-distance traveller; it has the longest migration of any animal species. Every year, this small tern flies from its wintering grounds in Antarctica, where it feeds mainly on fish and crustaceans, to its Arctic breeding grounds, where it feasts on the masses of insects. Depending on the route it takes, the tern covers approximately 35,000 kilometres by the time it reaches its destination; but on the upside, by travelling from one end of the world to the other and back every year, the bird experiences constant daylight. Tell me, sun worshippers, is that reward enough for such a tremendously lengthy journey? The bugs and fish certainly wouldn't be enough to entice me into such an undertaking.

No Rest for the Godwit

The Arctic tern may hold the record for the most distance travelled in migration, but the bar-tailed godwit takes the prize for the longest distance travelled without stopping. It should also get a prize for the biggest sacrifice made in the interests of migrating—to fuel its long journey, the bird's body destroys its internal organs. As it flies from its Arctic breeding grounds to spend the colder months in warmer climates such as New Zealand and Australia, the godwit's digestive tract shuts down and most of the bird's kidneys, gut and liver are destroyed, making the bird lighter and allowing metabolic energy to be focused where it is most needed, namely the heart and the muscles needed for flying. The organs regenerate once the bird has completed its journey, which can take eight or nine days.

Only eight or nine days, you say? Yup. This crazy bird is no laggard; when it migrates, it means business. The godwit undertakes its migration in one gruelling, nonstop journey. During this time, the bird never stops flapping its wings—using up more than eight times the amount of energy it consumes in its everyday activities, energy that will not be replaced until the bird reaches its destination. It doesn't eat or drink, and it doesn't sleep. To prepare itself for this journey, the godwit stuffs itself to pack on weight, increasing its mass by about half again and storing the excess calories as fat. By the time it departs on its migration, 55 percent of its body mass will be made up of fat, the highest amount of any bird species.

DID YOU **KNOW?**

In October 2008, a godwit fitted with a transmitter flew 11,680 kilometres from her Arctic breeding grounds to New Zealand in just over eight days—the longest nonstop journey ever recorded for any animal, regardless of species.

Generational Trail Blazers

Birds don't have the monopoly on impressive migrations. The monarch butterfly also undertakes an impressive journey, flying more than 4000 kilometres from its Mexican wintering grounds to its Canadian breeding grounds. The butterflies fly en masse, with millions of individuals flitting through the sky at one time. But the monarch's migration isn't remarkable because of the sheer number of participants; rather, it is noteworthy because the butterflies that make the migration south are the hatchlings of the batch that made the northern journey earlier in the year. In other words, the migrating butterflies undertaking this amazing journey have never done it before, and they have no guidance from experienced monarchs because members of the older generation that made the trip the other way died once they arrived.

To Warmer Climes

Migration is not limited only to those species that can fly—the humpback whale also undertakes a lengthy journey from its summer to its winter waters and back again. However, unlike most of our bird species, which come to Canada to breed, the whales do things a little backwards. They come to their northern grounds to stuff themselves on a cold-water buffet of krill and small fish before heading back south to breed.

Whales that spend their summers in Canadian waters do not all head to the same breeding grounds; humpbacks from the East Coast overwinter in the Caribbean (makes sense to me; who wouldn't want to spend the winter there?), whereas humpbacks from the Pacific head either to Japan or towards Mexico, depending on where exactly in the Pacific they spend their summers. Seems to me as though the East Coast whales are getting the better deal here.

Where's the Bug Spray When You Need It?

Herds of caribou have been migrating from their summer calving grounds in the Arctic tundra to their more southerly wintering grounds in Canada's coniferous forests since the time of mastodons and sabre-toothed cats. The round-trip journey can range from 2500 to 3000 kilometres and takes the animals across swollen, glacier-fed rivers, deep snow, slippery mountain slopes and countless other dangerous obstacles. When they arrive on their calving grounds, the pregnant females are in a sad state, the worst physical condition they face all year. And they only remain on their calving grounds for about two weeks before setting off again.

So why do they do it? Surprisingly, one of the main factors that drives caribou migration is insects. Nose bot and warble flies, in particular, are a huge source of irritation to caribou and can even put their health at risk. These large flies are relentless in their harassment, hovering in the caribou's faces and flying into their eyes and up their noses. Their antics panic the herd and have the caribou on the run to the point that the animals burn off most of their fat reserves, making them less able to survive winter. The flies also use caribou as incubators; warble fly eggs incubate in the caribou's flesh, with larvae emerging from its back in spring, and nose botfly larvae crawl into the ungulate's nose and continue their development in the unfortunate creature's nasal passages. Blech. Perhaps long-distance migration isn't such a bad thing, after all...

UNCONVENTIONAL DWELLINGS

A Good Shell Is Hard to Find

I don't know about you, but I sometimes think I would love to be able to just abandon my space and not have to worry about moving—just leave it all behind and find something more suitable. Perhaps I was a hermit crab in a past life. Although it is called a crab, the hermit crab is not closely related to true crabs. For one thing, it is missing the hard outer shell that we generally associate with crabs. The hermit crab's body has an exoskeleton that is strong enough to support it but not hard enough to protect its soft abdomen, so to protect itself from predators, the crab must improvise and find a shell it can move into. More often than not, the crab uses an abandoned sea snail shell, but there is not exactly a surplus of these shells lying around, at least not of a suitable size, so competition for those that are available can be fierce. A weird shell is still better than no shell, so in a pinch, the crab will use whatever it can find, including plastic bottle lids and those mini liquor bottles you get on airplanes. As the crab grows and its living quarters get a little tight, it repeats the whole process, abandoning its shell and searching for one that is more spacious. When the crab is stressed or feels threatened, it pulls its entire body into its salvaged home—a crab that doesn't have a well-fitted shell runs the risk of being eaten because it cannot withdraw completely out of the line of danger.

Waxy Wonderland

From the outside, a beehive may not look like much, but inside it is a complex structure made up of hexagonal cells built out of beeswax. Many bee species use a cavity, usually in a tree but sometimes also in rock, as the outer structure of their hive, then fill it in with honeycombs. The bees begin by smoothing down the outer bark around the hive's entrance. Once that is done, they move inside, coating the inner surface of the tree with a thin layer of propolis, a resinous substance that the bees gather from tree buds, filling in crevices in the wood. Now it is time to build the honeycombs. Parallel layers of honeycomb are attached to the inside of the tree, leaving enough space for bees to manoeuvre between the layers. The top layer is where the honey is kept, and beneath this are the pollen-storage layers. The worker bees occupy the layers directly under the pollen, and the drone bees get the layer under the workers. The very bottom of the hive is reserved for the queen bee.

Mobile Home

Snails have so many predators that it might be more appropriate to ask what doesn't feed on them. Not being particularly speedy, a snail can hardly make a dash to the safety of its home, outrunning would-be predators, so instead, it carries its home along wherever it goes. A snail is born with its shell, which is made of calcium carbonate, but a newly hatched snail's shell is soft and remains so until the little critter consumes some calcium. Upon hatching, the baby snail generally eats the shell of the egg it was born in, giving its own shell the calcium boost it needs to harden. As the snail grows, the shell grows along with it, and most snail species have shells that are whorled to the right. When they hibernate or estivate, snails close their shell with a mucous-like membrane, called an epiphragm, that helps seal in moisture while it keeps nasty weather, and even some predators, at bay.

> **estivation:** a state of dormancy much like hibernation that occurs in hot, dry weather

Safety First

Another creature that has sacrificed speed for safety is the turtle. Carrying your home around on your back has got to be slow, heavy work, but at least if you get tired, you can nap in relative safety wherever you like. A turtle's shell is actually the creature's modified rib cage and backbone. It is made up of the carapace, or upper shell, and the plastron, or lower shell. Despite what you may have seen in cartoons, the animal cannot crawl out of its shell—its body is firmly attached to the inside. As well, the turtle's epidermis grows over the outside of the shell in horny-looking scales, called scutes. The scutes are made of keratin and help to make the shell stronger. Most turtles have a high, rounded carapace so that predators cannot get their jaws around the shell, and, when threatened, most turtle species can fully withdraw their soft bits (limbs, head and so on) into the shell, out of reach of the frustrated predator.

Architect Extraordinaire

One of nature's most impressive and industrious architects, the beaver, is considered a keystone species. Its dam building doesn't just shape the landscape to fit its own needs—the wetlands that it creates support a wide variety of other species. Of course, the beaver doesn't really care about that; it just wants to build itself a comfy, secure home. The beaver's first step is to dam a stream so that it creates a pond large enough to build its lodge in. It amasses a bunch of branches, twigs and logs—sometimes even entire trees—and jams them into position. The dam's shape depends on the current of the water it is meant to block. Relatively still water can be held back by a straight dam, but dams in strong-flowing water need to be curved, with the centre pointing upstream.

Next comes the lodge. The beaver gnaws through the trunk of a tree, choosing one that is close to the water's edge, and once the tree has fallen over, the beaver drags the log to its lodge and wedges it into place. If the beaver has to use trees that are some distance from the water, it will build a canal before felling the tree to make transporting it to the lodge a much less strenuous task. Once the wooden structure is in place, the beaver covers the entire structure with mud, which freezes solid in winter and protects the family from any predators that might try to break in. The end result may not be pleasing to the eye, but it is an almost impenetrable fortress for the beaver family. The entrance to the lodge is underwater and opens onto the first of two main chambers, which is a sort of drip room, where the animal dries off before it moves into the other chamber, the main living area.

DID YOU  KNOW?

In 2007, a new record was set for the largest known beaver dam. Located in northern Alberta, the dam measured about 850 metres long.

Prairie Dog Metropolis

We all know that prairie dogs—often mistakenly called gophers, even though true gophers are in an entirely different family—build their homes underground, but the black-tailed prairie dog takes this to extremes. Found on (or should I say under?) the prairies in southern Saskatchewan, these prairie dogs live in massive colonies known as "towns." Each town is divided into "wards," the prairie dog equivalent of neighbourhoods, and wards are further divided into "coteries," which are the basic family units. All the animals in a coterie share blood ties. The average town covers about 100 hectares, but the largest town ever recorded was a whopping 64,750 square kilometres and included about 400 million prairie dogs, more than 13 times the human population of Canada (in fact, almost as much as the combined population of the U.S., Canada and Mexico!).

Wherever I Lay My Head…

The house wren is not picky about where it builds its home. Sure, tree cavities are the traditional choice for nesting sites, especially abandoned woodpecker cavities, but the house wrens of today are apparently not big on tradition. They will nest in almost anything. House wren nests have been found in old boots, flowerpots, tin cans, the pockets of laundry hung out to dry (which makes me wonder just how long the owner of the garments left them hanging on the line) and even in an old cow skull that was nailed to a wall.

All That Work for Nothing

It is the male marsh wren's responsibility to construct the nest that his family will be raised in, and he throws himself into the task with admirable zeal. The energetic little architect builds loads of dummy nests to attract his future mate, then he takes the female on a tour so she can inspect his handiwork and see if any are to her liking. If she likes what she sees, she sticks around, and he has found himself a mate. If she is not impressed, she flitters off to see what the next male has to offer.

Shelter Even from Above

It may seem to be nothing more than a messy heap of sticks and debris to us, but a magpie nest is actually an impressive work of engineering. The birds build the outer structure out of sticks and then glue the mass together with mud. They then set to work on the inside, making it all nice and cozy for the nestlings. The female lines the nest with fine vegetation such as roots, stems and grass, and any animal hair she can find is also thrown into the mix. Sounds pretty standard so far, right? Pretty much the same as any other bird nest?

Well, once the inside is fixed up just the way they like it, the magpies go one better than most other birds do—they build a roof.

Made of twigs—especially thorny ones, if possible—the roof has the dual purpose of providing shelter from the elements as well as acting as a deterrent for any would-be thieves looking to carry away an egg or a defenceless nestling. When it is finished, a magpie nest is so sturdy that even large birds of prey, such as owls, find it hard to break into; in fact, great horned owls are known to take over abandoned magpie nests to rear their own families. Imagine how strong and well built the nest must be to not only survive the battering winds and snows of winter but also to make it through in such good condition that it can still support the weight of a growing great horned owl family. You would think that after putting that kind of effort into building a nest, the magpies would reuse it the following year, but they seem to prefer to start fresh and build a new nest every breeding season, though they will sometimes remove the domed roof from a previous year's nest and incorporate it into their current building project. Scientists have suggested that magpies were once cavity nesters; however, because suitable cavities are in great demand but short supply, the birds switched to making their own "cavities" rather than relying on natural ones to become available.

LIKE A FISH OUT OF WATER

Look Out Below!

Puffins may be elegant as they swim underwater, and they even move about well on land, but they are not so good in flight. Horned puffins, found along the coast of BC, are powerful fliers, but they are somewhat lacking in control. As the birds circle above breeding colonies, individuals often collide with each other, sending the birds tumbling through the air until they can right themselves. Tufted puffins, also common along the northern Pacific coastline, fare a little better in the air—they usually manage to avoid crashing into each other—but their landings leave something to be desired. The wobbly birds often smash into rocks, talus or grassy slopes as they come in for a landing.

Less-than-Friendly Ground

Considered a symbol of the great Canadian wilderness, the common loon is a strong, steady flier, capable of reaching speeds of 120 kilometres per hour. However, its body is really designed for the water. With feet placed far back on its body, the loon is a graceful and agile swimmer and diver, fast and manoeuvrable as it pursues its fish prey under the water's surface. But the very adaptations that afford the bird such grace in the water do not work so well on land—the placement of its feet makes it nearly impossible for the loon to walk. If it must move about on land, the loon lays on its belly and uses both feet together to push itself while simultaneously using its "wrists" with half-opened wings to heave itself forward, dragging its underside across the ground. Loon chicks are able to walk about with ease on land for the first few weeks of their lives, but they begin to lose this mobility by the time they are three weeks old, and soon after, they must manoeuvre with the awkward shuffle used by their parents.

DID YOU KNOW?

Loons sometimes mistake sunbaked roads for bodies of water and land on the pavement. When this happens, they cannot get airborne again without human intervention. By the time these stranded loons are found, they have often damaged their feet and worn the wrist area of their wings down to the bone trying to move about on the hard ground.

A Fish-like Mammal Out of Water

The blue whale can grow so astoundingly large because it lives in water—the salt water supports its incredible bulk, removing the need for the sturdy, weight-bearing bones that a similar-sized mammal living on land would need. However, the lack of a strong bone structure can some-times be the whale's undoing; if the creature were to beach itself or get trapped in shallow water, the weight of its body unsupported by water would crush its lungs, and it would suffocate.

It's Getting Hot Out Here

The northern fur seal's thick coat keeps the creature warm in the frigid waters of the North Pacific (and is probably the envy of marine mammals and perhaps even a few birds throughout the region), but it is not so handy once the seal is out of the water. The well-insulated animal quickly over-heats when on land, especially if the sun is out, and can die of heat stroke. Northern fur seals hang out along the BC coast and come ashore only to breed, but because they overheat so easily and quickly, there are few places where they can stay on land long enough to birth and rear their pups. The seals will travel up to 10,000 kilometres to reach their northern breeding grounds, which takes them out of Canadian waters and onto Alaskan and Siberian coastlines, where weather conditions are often overcast and misty. Young male fur seals do not come ashore for the first few years of their lives, sometimes not before they are five or so years of age and big enough to compete with the large, dominant bulls for harems. Why risk heatstroke when you are too small to breed?

URBAN ANIMALS

Furry Urban Burglar

Many a homeowner has cursed the urban raccoon. With its nimble fingers and keen intelligence, the raccoon has proven to be a formidable foe for people trying to keep their garbage unmolested until pickup day. The raccoon is opportunistic and will eat pretty much anything it finds. Human garbage is like a smorgasbord for this clever creature. It seems that no matter what steps a homeowner takes to keep it out, the raccoon can always do one better. If you put a lid on your garbage can, the raccoon will remove it. If you fasten the lid in place with string, the raccoon will chew through it or untie it. Put a gate between your garbage can and the raccoon, and the sneaky animal will unlatch it. People who have tried using wildlife-proof containers have woken to find that the persistent little bandit has made off with the entire bin. Not only does the raccoon raid our garbage cans, it also makes itself at home in our attics, helps itself to our pets' food and snacks at its leisure from our gardens. There seems to be no limit to what this bold creature will do.

Just Lounging Around Town

Anyone who wishes to see an elk, also known as a wapiti, needs only visit Jasper National Park. And I don't mean you should go slogging through the backcountry in the hope of seeing one of these large ungulates—just take a simple stroll down the main street in Jasper townsite at dusk, and you will almost assuredly have a variety of individuals to feast your eyes on. The elk, clever creatures that they are, have figured out that not only does the town have

plenty of easily accessible food (such as people's lawns and gardens), it is also predator free. The wolves and mountain lions that would normally plague the elk shy away from people and will not venture into town in pursuit of their prey. So the elk laze about on the street meridians and in the schoolyard, contentedly munching on grass.

Break-and-Enter Bears

People living in bear country know what steps they should take so as not to attract bears to their property. They are told not to leave their garbage lying on the curb, to keep their pet food inside and to scrub the meat residue off their barbeque grills. But for the past few years in BC, bears have not been content with licking people's barbeque grills or tearing up their garbage—the bears are letting themselves into people's homes to search for food. Bears have been climbing through partially open windows and opening unlocked doors, pushing down handles with their paws or turning the knobs with their teeth. They search through homeowners' cupboards, help themselves to food left on counters or tables and even raid refrigerators. Bears have also been breaking into sheds and vehicles. In North Vancouver and West Vancouver, the spoils have been so good that some bears are even delaying their hibernation or skipping it all together. Why waste time sleeping when you could be snacking?

MUTUALLY BENEFICIAL

Whale Bird

The relationship between the red phalarope and several whale species is so well established that whale watchers often keep an eye out for the shorebird to find out where the whales are. The phalaropes feed on crustaceans that are brought to the surface when grey or bowhead whales, to name a few, come up for air. The bold bird has also been seen landing on the whales' backs to pick parasites off their skin.

Dangerous Liaisons

The relationship between aphids and the ants that protect them may, at first glance, seem to be a win-win association. After all, the aphids get to go about their business confident that they are safe from being eaten by one of their many predators, and the ants get to feast on the honeydew the aphids produce. But upon closer inspection, the relationship is not as mutualistic as it first appears—and it seems as though the aphids are getting the raw deal. First, the ants control the aphids with chemicals on their feet that act as tranquilizers, preventing the aphids from moving on to greener pastures, so to speak. Sometimes the ants resort to a harsher form of control—they bite the aphids' wings off so that they cannot escape. And from time to time, a hungry ant will snack on one of the aphid workers. Perhaps instead of thinking of the ant-aphid relationship as one of cooperation, we should visualize it as ant farmers managing little aphid ranches.

Follow That Bird

An unopened carcass is of no use to a raven. There are only a few animals that can open a carcass, and the raven is certainly not one of them—it cannot break through the animal's tough skin to get at the goodies inside. So the raven, clever bird that it is, enlists the help of other carnivores. It knows that large predators such as wolves, bears or cougars can easily tear through a dead animal's thick hide, so it attracts a larger animal's attention, leads it to the carcass, waits for it to do its work, then sneaks in and steals as many scraps as it can. If the predator isn't in a sharing mood, the bird might wait until the larger animal has eaten its fill before heading over to grab a few tidbits. The raven has even been known to lead predators to prey animals (ones that are still alive and able-bodied), presumably in

the hope that the predator will be able to take the prey down and the bird can then sneak a meal. Why hunt when you can have someone else do all the hard work for you? Anecdotal evidence also credits the bird with leading human hunters to animals such as deer or elk.

Nothing Beats Body Heat

Raccoons are known as solitary creatures, but in the colder parts of their range (like, say, Canada) they have been known to bunk together to weather cold winters. By huddling together, they can stay warmer while using less energy, which is critical because, though not true hibernators, these critters are not active during cold or snowy weather. Instead, they rely on their fat reserves. As many as 23 raccoons have been found sharing one den. The striped skunk, also a solitary creature, will do the same thing in the colder regions of its habitat—groups of as many as 20 skunks have been found sharing a den.

A Loose Alliance

They may not be the best of pals, but coyotes and badgers will work together to catch a meal. The coyote follows the hunting badger, catching rodents that flee their burrows through alternate holes, while the badger digs relentlessly at the main opening. In return, the coyote will lead the badger to new burrows or hunting areas. There is no evidence that the animals share the spoils (basically, whoever catches the prey gets to eat it), but they do work cooperatively (though you could argue that the coyote is getting the better end of the deal, since the badger has to do all the digging and therefore most of the hard work).

A Little Respect, Please

Yes, most of us love to hate them, but lowly ring-billed gulls actually do us a great service. As these birds flock behind farm machinery, they eat the insects that are disturbed when the machinery passes, thereby eliminating loads of the nasty critters that plague farmers. This saves the farmers money that they would otherwise have to spend on insect control, and it prevents a great deal of crop damage. The same could be said for another much-maligned creature—the striped skunk. The smelly little critters eat many agricultural pests, including armyworms, potato beetles, snails and even rats and mice.

Befriending Bovines

The cattle egret catches about 95 percent of its food thanks to the hoofs of grazing cattle that stir up bugs from the grass. These egrets eat mostly grasshoppers, crickets and flies that linger around cattle. They generally run around the bovines' legs, but will also perch on the animals' backs. The cattle don't seem to mind. The egrets eat a lot of the flies that would otherwise be buzzing about, dive-bombing the animals' faces and irritating sensitive eyes and ears. During the breeding season, adult egrets feeding their young catch on average 638 prey items each day. In their native Africa, these egrets perform a similar service for many different species, including elephants, rhinos and hippopotamuses.

A BIT MORE ONE-SIDED

I'll Follow Your Lead

Black vultures have a poorly developed sense of smell, which seems like a bit of an evolutionary oversight. I mean, what better way to find a rotting carcass than by sniffing it out?

Instead of a strong sniffer, evolution has provided these birds with sharp eyesight, so they have had to be somewhat creative in developing a strategy to use it to their best advantage. Black vultures use their keen eyesight to watch for turkey vultures—birds that do have a keen sense of smell. Once the turkey vulture spots a carcass, or should I say, once a black vulture sees a turkey vulture spot a carcass, it follows the turkey vulture to the meal and often usurps the find. Because they rely on turkey vultures to find food, black vultures fly much higher than turkey vultures, reaching heights of more than 300 metres.

The Dark Side of Ants

Ants are hard workers, right? Isn't that the lesson we are taught with the whole grasshopper and ant story—the grasshopper lazes about while the ant works feverishly, and when winter comes, the grasshopper has nothing and the ant is sitting pretty. In our society, ants are held up as paragons of hard work and organization—organization may be true, but some ant species clearly didn't get the hard-work memo (though I suppose you could argue that pillaging, kidnapping and slavery are hard work).

Harpagoxenus canadensis, a species of the Formica family of ants, raids the nests of other ant species, stealing the eggs and bringing them back to their own nest, where the young hatch into a life of slavery. The youngsters, not knowing any better, will do all the grunt work in their new home and will even defend it from attack by their own species, should the need arise.

Unintentional Environmentalists

Thanks to their playful nature and cute, fuzzy, little faces, sea otters are among the most popular mammal species. What many people don't realize, though, is that we humans owe the sea otter a debt of gratitude, not just because they entertain us, but also because they help preserve our coastlines. Sea otters love to eat sea urchins; sea urchins love to eat kelp. If sea urchins feed too heavily on kelp, the kelp beds suffer and cannot buffer the coast against the eroding force of the Pacific tides. Without the protection offered by the kelp beds, exposed areas along the coast could be flooded or washed away entirely. Thankfully, sea otters are helping to save the coast from such a fate. The otters eat the urchins so that the urchins don't eat the kelp beds; the kelp beds continue to flourish and our coastlines remain intact—everyone's happy. Well, except maybe the sea urchins…

Enter At Your Own Risk

The Arctic lion's mane jellyfish is the largest known jellyfish in all the world's oceans. It is found in cold Pacific and Atlantic waters. The largest specimen of this creature that has been found to date was even longer than a blue whale when its long, trailing tentacles were included in the measurement. In the open ocean, these tentacles make a handy hiding spot for shrimp and many species of fish. Not only are the smaller creatures protected from predators, the tentacles also provide a constant supply of food.

Four Limbs Are Plenty

When's the last time you saw a five-legged frog hop by? Well, okay, I've never actually seen one, but they are out there! Many frog species, especially the Pacific treefrog, are victims of the trematode flatworm. This parasitic species depends on a number of host species for its development. First, the eggs develop in the body of a heron. When the bird relieves itself in the water, its trematode-egg-laden feces are often eaten by snails. The eggs hatch within the snail and eventually dig their way out of its body, heading directly for the nearest tadpole and digging into its limb buds. As the tadpole develops into a frog, the trematode disrupts the natural animal's growth—it can either prevent the limb bud from developing into a proper leg, resulting in a frog that is missing a limb or two, or it can cause the limb bud to duplicate, resulting in a frog with extra limbs, sometimes as many as 10.

UNEXPECTED ASSOCIATIONS

Hare vs. Muskox?

It is strange to think that the lowly, harmless-looking Arctic hare could be responsible for the death of large animals such as caribou and the impressive muskox, but it is true. In the harsh environment that the three species share, food is scarce. All three species feed on lichen and moss, as well as other plants, and food that the hare eats means that much less for the other species.

Medical Maggots

Maggots, especially the larvae of blowflies, are being put to work. These insects feed only on dead and damaged tissue, so doctors are using them to treat burn victims or people coming out of surgery. Maggots are placed on the wound, and the area is wrapped in a bandage. After two to three days, the maggots are removed. In the meantime, they will have eaten all the dead tissue as well as any bacteria that might have been hanging around. To feed, the maggots secrete digestive juices that dissolve the tissue, then basically lick or suck up the liquefied mess. People treated in this manner experience fewer complications from infection, and there is even evidence that it may help reduce scarring.

Creating Homes for Everyone

The industrious beaver is well known for its landscape-altering capabilities, but it now seems that this busy rodent is doing even more good then previously realized. By modifying its habitat, the beaver is helping to prop up struggling frog populations. Many amphibian species have been in decline worldwide for the past 30 or so years; today, more than one-third of amphibian species worldwide are staring extinction in the face, but the beaver may help reverse this situation in Canada. The beaver's changes to the landscape create environments that are favourable to tadpole survival—the presence of the dams doesn't just make the water more still, it also increases the water temperature and makes it rich in nutrients.

Researchers in Alberta noticed that populations of adult wood frogs, boreal chorus frogs and western toads (a Red List species according to COSEWIC) were higher in areas that had beaver dams. As well, when the stream water was tested, researchers found six times more young wood frogs, 24 times more young boreal frogs and 29 times as many young western toads as were found in streams that were not dammed by beavers. Researchers are now looking into ways to boost the beaver population (and therefore, hopefully, the frog and toad population) without causing conflict with human interests.

FURTHER FACTOIDS

No Match for Metal

Unfortunately, skunks are very often killed on our nation's roads. They wander along the roadways, snacking on bugs that have been squashed by vehicles. The little stinkers have so much confidence in their smelly weaponry and are so used to predators backing down that they stand their ground whenever threatened, regardless of the foe—not a good survival strategy when they are facing tonnes of fast-moving steel.

Directional Shedding

Like many mammals, the short-tailed shrew sheds its coat with the seasons, losing its summer pelage in late autumn and replacing it with a longer, thicker winter coat. The odd thing, though, is that in autumn, when the winter coat is coming in, the moult begins at the tail and works its way up the body, but in spring, when the old coat is shed and the summer coat comes in, the moult starts at the animal's head and moves downwards to the tail. No one knows why.

pelage: a mammal's coat; can be fur, hair or wool

The Hoof's on the Other Foot

Male caribou sport huge racks of antlers that can reach 1.5 metres in length; the females also have antlers, but they are much smaller. In caribou social structure, as with most members of the deer family, the buck with the largest antlers is dominant in the herd. However, dominant males should think carefully before they start throwing their weight around, at least where the females are concerned, because come winter, the hoof is on the other foot, so to speak. Male caribou shed their antlers in winter, but the females keep theirs, and for a short time, the females become dominant over the antler-less males.

Ghost of the Rainforest

If it's white and it's a bear, it must be a polar bear, right? Wrong. The Kermode bear, found on Princess Royal Island, British Columbia, is a white race of black bear. The colour of its coat is the result of a recessive gene, and white bears tend to choose other white bears to breed with, keeping the white population alive. This bear holds a special place in the mythology of some of the Native cultures that live along the BC coast.

Keep Your Distance

The Ord's kangaroo rat lives a solitary lifestyle and, like any hermit, prefers it that way. It may look sweet and cuddly, but this little kangaroo rat is fiercely territorial and has no qualms about fighting dirty to drive trespassers off its home range. When another animal gets too close for the kangaroo rat's comfort, the antisocial creature flies into action, boxing the invader with its little front paws, kicking with its much stronger back legs and—adding insult to potential injury—using its giant clown feet to kick sand into the intruder's face.

DID YOU KNOW?

In a single leap, the Ord's kangaroo rat can jump more than 10 times the length of its own body. Like its Australian namesake, the kangaroo rat gets around mostly by hopping on its back legs. It only uses its front legs when it is walking slowly, such as when it is searching for food.

Take Two Beavers and Call Me in the Morning

As do many mammal species, beavers use scent to mark their territory. However, unlike other mammals, beavers use a substance called castoreum in their scent marking. Castoreum is unique to beavers—it is produced in "castor sacs," glands connected to the beaver's urinary tract and located under the skin at the base of the tail, and one of its primary compounds is salicin. Beavers feed on the bark and small branches of willow trees, which contain salicin. When the beavers digest the wood, salicin from the willow is stored in their bodies, where it is used in the production of castoreum. As well as giving castoreum its unique scent, salicin has medicinal properties. Willow trees have been used to manage pain since the time of Hippocrates, and salicin was originally one of the main ingredients in aspirin. Castoreum, too, had a place in the medicine cabinets of yesteryear. At one time, the yellowish substance was taken to relieve minor aches and pains, as well as the discomfort associated with ulcers. Today it is used mostly in the perfume industry, but it still has a place in homeopathy, where it is recommended as a remedy for hysteria and restless, nightmare-laced sleep. It has even been used as an ingredient in cigarettes to give the tobacco a "sweet odour and smoky flavour."

Special Spew

Having trouble trying to decide what to get for that person who has everything? How about a nice chunk of petrified sperm whale vomit, otherwise known as ambergris? It may sound like a disgusting gift, but it is apparently much in demand, and because it is rare and difficult to acquire (it can only be found bobbing about on the ocean's surface or, occasionally, washed up on shore), it is very costly.

Ambergris starts out in the stomach of a sperm whale as a waxy substance that coats sharp, indigestible bits of the whale's diet—such as the beaks of giant squid—so that they do not rip or get caught in the whale's stomach or intestines. Once the sharp bits are coated in the ambergris, they can pass smoothly from the whale's body with no risk to the animal. After it is expelled from the sperm whale's stomach (and it can come out of either end, if you get my meaning), the smelly mess floats on the ocean's surface, where it hardens and, apparently, undergoes an olfactory transition from putrid to perfumy.

Ambergris was once used in perfumes, not only for its unique scent, but also because it gave the perfume's scent extra staying power, keeping it on the skin instead of evaporating. Ambergris was also used in medicine and, apparently, as a seasoning for food. There really is no accounting for taste.

Many countries won't allow the sale of ambergris because sperm whales are endangered, and therefore no sperm whale products can be sold. So, depending on what country the stuff washes up in, or which waters it is pulled from, it may or may not be legal to possess it. Synthetic ambergris has mostly replaced the real thing in the perfume industry. Ambergris can be white, grey, black or a combination of all three colours.

Magic Bullet

What used to be one of our more common frogs may one day soon help cure people suffering from brain tumours. The northern leopard frog, found throughout most of Canada (but in decline in some provinces, such as BC), has a substance called amphinase that occurs in the frog's egg cells. Amphinase attacks and kills tumour cells because the tumour cells have a sugary coating that the amphinase molecule targets. The amphinase binds to the cell, then penetrates it, inactivating the cell's ribonucleic acid (RNA) and killing the tumour. As Ravi Acharya, one of the British scientists responsible for the discovery, puts it, amphinase is "rather like Mother Nature's very own magic bullet for recognizing and destroying cancer cells." Scientists have high hopes that the molecule can be synthesized and used as a therapeutic treatment for humans.

Silky but Strong

What is thin, thread-like and stretchy but has a tensile strength greater than that of steel? Give up? The answer is spider silk. This strange substance is discharged from the spinnerets located on the spider's abdomen. To start building its web, the spider shoots a line of silk into the wind, letting the breeze carry the thread to a surface that it can adhere to. Once that first strand has been placed,

the spider can use it as a base to weave the rest of the complex lattice. Because of its strength and elasticity, researchers are looking into developing a similar substance that can be used in bulletproof vests.

DID YOU KNOW?

Spiders do not stick to their own webs because they weave both adhesive and non-adhesive threads into their lattice, and they are careful to move about only on the non-adhesive threads.

Toxic Poo

If you were to visit a bathroom that had just been vacated by a person who ate a particularly bad burrito, you might find it easy to believe that stool could be toxic, but in the case of the northern fulmar, it actually is. Fulmar poo has high levels of toxins such as mercury and DDT; the pollutants make their way to the Arctic thanks to long-range transport from southern regions. Because seabirds are one of the top predators in the marine ecosystem, chemicals and toxins accumulate in their bodies from the foods they consume. Fulmars are also long-lived birds, sometimes reaching more than 50 years of age, so they have a lot of opportunity to bioaccumulate toxins in their fatty tissues. In fact, these birds are believed to bioaccumulate more toxins and pollutants than other seabirds that share their habitat. This may be because these seabirds often feed on sea mammal carcasses, especially the blubber of whales and walruses, which also stores contaminants.

When the fulmars' bodies start burning off the fat reserves, such as during breeding or if food is scarce, the toxins stored in the fat are mobilized, and the birds then pass some of these contaminants into the environment through their droppings. Basically, the fulmars are taking the contaminants from the water and passing them onto the land, spreading the contamination. The toxins are so concentrated at some fulmar breeding sites that fishermen are advised to keep away from them.

It Just Makes Good Sense

There are many good reasons why birds' eggs are oval in shape. First, by being oval, the eggs will roll in a circle rather than a straight line if disturbed. This way, the egg has less of a chance of rolling out of the nest and smashing onto the ground below—it simply rolls around itself, not covering any real distance. The shape also makes it easy for more eggs to fit in a nest with less wasted space—if they were circular, there would be more space between them where cold air could touch the egg's surface. Also, the oval shape allows the eggs to nestle closer together, taking up less room in cramped quarters.

Predator in Pursuit

Ever wonder why animals such as rabbits hop madly down the road in front of your car instead of jumping immediately to the side? The answer has to do with their predator evasion instinct. Prey animals attempting to outrun a pursuer try to keep the animal directly behind them. A predator generally attacks its prey around the throat or neck area, so prey species keep their less-vulnerable bottoms pointing towards the would-be attacker. To throw off the animal in pursuit, the prey species usually runs in a straight line, then dives suddenly to the side, hoping to catch the predator by surprise so that the pursuer overshoots it.

Water Shy

For such aquatic animals, otters are a little slow to take to the water. Otter babies are introduced to the water by their mothers when they are about seven weeks old, but they do not always go quietly—some must be dragged unceremoniously into the water by the scruff of the neck.

It's Getting Hot in Here

If you've noticed that the winters in your area are a little milder of late, blame it on the livestock industry; cows may be responsible for up to 20 percent of global methane emissions. One cow can be responsible for more VOCs (volatile organic compounds—environmental nasties) than a small car. Sheep, too, contribute their fair share. To be fair, it is not only cattle and sheep that are responsible for elevated methane levels. Termites also share the blame; they are thought to be responsible for four percent of global methane emissions. The gas builds up in their hind-guts as a result of their woody diet, and then oozes out of their bodies.

Brainy Whale

The sperm whale has the largest brain of any animal, but not relative to body size. The brain makes up only about 0.02 percent of the whale's overall body weight.

Now That's Just Unmannerly!

We have yet another reason to hate mosquitoes. Not only do they interrupt our rest by filling our ears with their annoying, high-pitched whine as we are trying to sleep, rob us of our blood and then leave behind itchy welts, they also pee on us. After biting us and drinking their fill of our blood, females are so engorged that they cannot take flight. To reduce their body weight so that it is more manageable in the air, the females pump out all the excess liquid from the blood they have consumed, getting rid of any substance that is not useful to them. They have to do this before they can fly, so they leave the excreted substance (that is, pee) on our skin. Hmph! Never again will I will feel guilty for squashing one of the little monsters.

UNDER SIEGE!

Alien Nation

We've been invaded! Aliens are running amok throughout the country! True, they are not the kind that typically feature in Hollywood blockbusters, but they can be pretty scary just the same. Alien species are those that have moved out of their native habitats, usually because of humans, and into areas where they do not occur naturally. Sometimes they are brought into the country intentionally, and sometimes they sneak their way across our borders, without us even being aware of their presence. Well, not initially, at least. We soon cotton on to their existence, though, when they muck about with our own native species. Because they are out of their natural habitat, many alien species have a competitive edge over the species native to the area; they often do not have natural predators in their new home, are not affected by the local parasites and are immune to diseases prevalent in the region. Alien species can have a huge impact on indigenous species, either directly (by preying on them) or indirectly (by outcompeting them for denning and nesting sites or food).

Unwelcome Guests

So how do these alien species make it into Canada? A few species made it to our country entirely on their own, and their presence here is a reflection of an expansion of their traditional range. Some were brought in intentionally by what I like to think of as scandalously shortsighted people who were trying to manage one biological problem by introducing another. Others snuck across the border, hidden in packaging or wooden shipping crates used to transport imported goods. Many marine species arrived in the ballast water of ships, and some insects were brought in along with welcome imports such as plants used in landscaping, with the importer never realizing the unwelcome hitchhiker was along for the ride. At one time, most of the alien species coming into Canada arrived from Europe, but recently, the number of species coming in from Asia has been on the rise.

UNINVITED IMPORTS

Sea Lamprey

Now this is one nasty creature—it literally sucks the life out of its prey. Native to the Mediterranean and North Atlantic coastlines, the sea lamprey has moved into the Great Lakes.

Lampreys have existed in Lake Ontario for ages but did not spread into the other lakes until the early 1800s. This jawless, boneless, scaleless fish is responsible for the extinction of three species of whitefish in the Great Lakes. It has a mouth that is shaped like a toothy "O," which it attaches to the bodies of healthy fish to suck out the unfortunate creatures' blood and then the tissue. Eventually the lamprey will extract everything from the fish's body except the bones.

House Sparrow

Originally from Europe, the house sparrow now lives on all continents except Antarctica. This feisty little bird is not actually a sparrow; it is more closely related to the Old World weaver finches. The house sparrow is an amazingly adaptable bird, outcompeting many of our native birds for nesting sites. It was introduced to North America between 1850 and 1875 by some misguided people who thought it would help control insect pests. Although it is thriving in North America, its numbers have declined drastically throughout its range in Europe.

Cattle Egret

You can't help but be impressed with this little bird, and not only because it looks so cool as it struts through the fields at the feet of cows, with its spiky Lisa Simpson–like hairdo blowing in the wind. The cattle egret is originally from Africa, but it has managed to expand its range with no assistance from humans. Today, this egret lives on every continent except Antarctica. To leave its native home, the cattle egret had to make a 3060-kilometre ocean crossing from the mainland of Africa to the tip of South America before it could land on solid ground once more. So why did it leave? No one knows, but one theory proposes that a population of cattle egrets got blown off course by some strong winds, and they just kept going. The first North American sightings were made in the 1940s.

Rock Pigeon

Pigeons are so common and widespread in Canadian cities that it is actually seems weird to go through a day without seeing one. These charismatic birds are originally from Europe, northern Africa and Asia, and they were brought into Canada in the mid-1800s by colonists arriving with their domesticated birds. It has been suggested that the first pigeons brought to North America arrived in Port Royal, Nova Scotia. Rock pigeons are one of the few birds in Canada that can breed at any time of the year.

Norway Rat

It's no wonder these creatures are colonizing the world. They are practically indestructible! As proof, I offer the following information: a rat can fall from a height of 15 metres and walk away none the worse for the experience; a rat can tread water for up to three days; and a rat can survive being flushed down the toilet (I'm not sure how that little tidbit came to be known, and I hope it and the previous point are in no way related). It can even turn around and climb back up the pipes to gain re-entry into the building from which it was so rudely expelled. Rats came to Canada by way of Europe, but they are not native to that continent, either (nor are they welcome there, what with being at least partially responsible for wiping out a large portion of the population in medieval times). The rat's origins have been traced to Asia, perhaps northern China.

DID YOU KNOW?

If you want to live in a rat-free area, you can head to either Alberta or Antarctica. They are the only two regions in the world that the Norway rat has not been able to colonize.

Coyote in Newfoundland

I know what you're thinking. How can a species be an alien in its native country? Well, coyotes are not indigenous to the island of Newfoundland. These canines have been expanding their range across North America for decades. Originally an animal of the prairies, the coyote has been steadily spreading throughout the country, taking advantage of human activities such as agriculture and filling in a niche left behind by the widespread eradication of wolves at the hands of farmers and ranchers. It is one thing for the canine to expand across the mainland of Canada, but for it to trudge on and settle on the island of Newfoundland is a bit strange. Although reported sightings of the crafty canine began trickling in around 1985, the coyote's presence on the Rock wasn't confirmed until 1987, when a coyote pup crossing the road was struck and killed by a car. So how did they make their way from the mainland to the island? Most likely, they trotted across ice bridges from the coasts of Nova Scotia or Labrador.

Asian Long-horned Beetle

These rather cool-looking beetles found their way from China, Korea and Japan into North America in shipping crates made from untreated wood, as well as other forms of packaging. Once they emerged from their hidey-holes, the beetles continued their spread throughout the country. They have even been seen hitching rides on moving vehicles (lazy critters). The larvae are destructive to some of our native trees, including poplar, elm and maple. They attack the inner wood, disrupting sap flow and creating tunnels throughout the trunk, essentially strangling the tree from the inside.

Zebra Mussel

Seafood lovers may like nothing better than a steaming plate of mussels, but I think even the most devoted enthusiast would think twice before digging into these babies. Originally from Eurasia, particularly the Caspian, Black and Asov seas, zebra mussels hitched a ride to the Great Lakes in ballast water in the 1980s. They are now widespread throughout the lakes system, where they clog waterways and plug industrial intake and output pipes, absorbing many of the contaminants that travel down the tubes. The Canadian and American governments have spent astronomical amounts of money trying to rid the Great Lakes of this nuisance species, but the mussel is most likely here to stay.

UNEXPECTED GUESTS

That's Quite a Beak

The black skimmer's unique bill may look like a case of cosmetic surgery gone wrong, but it is actually a handy adaptation to the bird's feeding style. The skimmer, as its name implies, flies close to the water's surface while searching for something to snack on. As it flies, the bird holds its dagger-like bill open, running the lower portion, which can be 2 to 3 centimetres longer than the upper mandible, just under the surface of the water. When the bill comes into contact with a fish, it snaps shut. The black skimmer is not native to Canada, but it has been spotted on occasion in PEI.

Gone Astray

Although they are most commonly seen in tropical waters, magnificent frigatebirds have a wide range along the Atlantic and Pacific coasts, and these birds really get around. Vagrants have been seen as far from home as Alaska. In Canada, magnificent frigatebirds have been spotted in Ontario and Québec, and one tired bird hitched a ride into a BC harbour on the deck of a freighter. These birds are reputed to ride in the winds that precede hurricanes, so it is no wonder they can get blown a little off course.

Shark Attack

Is it just me that finds the presence of great white sharks in Canadian waters bizarre? Normally the words "great white" bring to mind California beaches or South African waters, not the rather chilly waters of the coasts of Newfoundland, Prince Edward Island and British Columbia. Nevertheless, they're here. Great whites can be found in all the world's oceans, though they are rare in Atlantic Canada. These sharks usually stick to coastal or offshore waters around continental shelves, but they also stray into shallower water such as bays and harbours and can be found in the water column from surface waters down to about 1200 metres. They are the world's largest predatory fish, reaching lengths of six metres, and their mouths contain more than 3000 serrated teeth. Think about that the next time you head to the beach for a little dip.

Sniff and Ye Shall Find

Many birds have their own smell, which is not always pleasing to the human nose (try walking past an egret rookery sometime), but one bird you would be happy to smell is the crested auklet. These sooty grey seabirds smell like ripe tangerines. Researchers have suggested several theories to explain the smell. It may be a pheromone used in courtship, or it may be a way that the birds can find one another when they cannot rely on their eyesight or hearing. Crested auklets spend most of their time in the Bering Sea, where fog often shrouds the landscape and the crashing of waves drowns out all but the loudest of noises. In that type of environment, a distinctive smell might be the difference between a bird that finds its way back to the breeding colony and one that overshoots its mark. Hmm...Maybe the vagrant auklets that are occasionally spotted in BC are birds with stuffy noses.

IS ANYONE OUT THERE?

Watching for Sasquatch

Could there really be a giant, ape-like creature wandering through the forests of British Columbia? Some people think so. The sasquatch is represented in the folklore of Native cultures along the BC coast (although it differs in form and habit depending on the culture), and sightings by Europeans date as far back as the mid-1800s. This creature is said to make its home in the wilderness of the Pacific Northwest, but reports of sightings have also trickled in from Alberta, Saskatchewan and even as far east as Ontario.

The sasquatch has been described as a massive creature ranging from 2 to 4.5 metres tall and weighing in at more than 230 kilograms (though where these numbers come from I couldn't tell you, since a sasquatch body has never been found). Its fur is believed to be rather bear-like, ranging from dark brown to cinnamon. The creature is said to be bipedal, standing upright but a little hunched, and it apparently has a rather gorilla-like face, complete with a heavy brow ridge and sagittal crest. People who have reported seeing the sasquatch describe it as an incredibly foul-smelling creature—its scent has been described as a combination of skunk spray and wet dog. Nice. The fact that there are so few sightings has been explained by suggesting that the cryptid is nocturnal by nature.

Some suggest that sasquatch sightings are actually misidentified bears, but there is another school of thought that believes the mysterious animal may represent a remnant population of *Gigantopithecus*, a creature that lived more than 300,000 years ago in China and India. So do we have sasquatch in Canada? To date, the sum of the "proof" that supports its existence includes an assortment of footprint casts (some of which have been proven to be the result of hoaxes and others that may actually be of a grizzly's foot) and a whole lot of anecdotal evidence. But hey, no one believed the world was round, either, until scientific evidence proved it to be so.

> **cryptid:** an animal whose existence has not been scientifically proven, such as the sasquatch
>
> **sagittal crest:** a ridge of bone that runs along the top of the skull and to which the jaw muscles are attached

Part Bear, Part Dog, All Trouble

According to the legends of some Native American cultures, a giant, wolf-like creature prowls the borderlands between Alaska and the Northwest Territories. Sightings of this fierce canine, known as the *waheela*, were especially common in the Nahanni Valley (also known as Headless Valley, thanks to the nasty pooch) and still trickle in from the area today. The waheela is described as a massive, stocky wolf with a long, snow-white coat. Its back legs are said to be shorter than its front ones, and its huge paws have widely spaced toes. As for its personality, the waheela seems to be antisocial. It shuns the pack life preferred by true wolves and lives alone in remote, largely inhospitable areas, stalking and beheading anyone foolish enough to wander into its territory. Native Americans believed the wolf was an evil spirit, but cryptozoologists that have taken an interest in the waheela suggest that it might be a modern-day version of the pre-historic bear-dog or dire wolf. I like to think that it is the sasquatch's faithful companion.

> **cryptozoologist:** a person who studies animals that have not been scientifically proven to exist

Friend or Foe?

Next time you are paddling your canoe in Lake Okanagan, keep an eye out for the Ogopogo. It should be pretty hard to miss, if it's around. Just look for a creature that is between 6 and 15 metres long, with a horse-shaped head and the body of a sea serpent. It may be dark blue, black or brown—reports vary. And keep your video camera ready so you can catch the creature on film; Ogopogo supporters will be most grateful for solid documentation that their beloved lake monster actually exists. Native peoples living

around Lake Okanagan before European settlers arrived in the area referred to the creature as N'ha-a-itk, or "lake demon." They believed it lived in an underwater cave under Squally Point, in the Peachland area, and they would not pass its lair without offering a gift of food lest it should decide to attack and eat them. The first recorded sighting by a Caucasian dates back to 1872, and there has been a rather steady stream of sightings ever since (though modern-day accounts portray the creature as friendly, not aggressive). Ogopogo has been compared to the basilosaurus, a prehistoric whale from the Eocene epoch with a long, snake-like body.

Champy
British Columbia isn't the only province that boasts a lake monster. Apparently Québec is also getting in on the action. The Lake Champlain Monster, a.k.a. Champ or Champy, is described as a plesiosaur-like creature with horns. The beast featured in the folklore of the Iroquois who settled in the region, and more than 300 sightings have been reported

over the past 300 or so years by European settlers and their descendents. The most recent Champy sighting occurred in 2005, when some fishermen caught what they believed to be the creature on video. Lake Champlain, which spans the border between Québec and the U.S., is about 121 metres deep in places, and it stretches over 175 kilometres long and 17 kilometres wide—more than enough space to support a number of large fish and other marine animals. But large enough to support a supposedly extinct dinosaur? Who knows.

STRANGE HYBRIDS

Gone to the Dogs

It doesn't happen very often, but from time to time a coyote and a domestic dog get a little too friendly. The resulting offspring are called coydogs if the father was the coyote and the mother was the domestic dog, or dogotes if the

father was the dog and the mother was the coyote (seriously, couldn't scientists come up with better names for these creatures?). These hybrids are rare because coyotes and domestic dogs generally view each other with hostility, not as potential mates. Coydogs and dogotes are thought to be more aggressive and less wary of humans than true coyotes.

Hard to Bear

In 2006, an American hunter looking to bag himself a polar bear trophy in the Northwest Territories got more than he bargained for when he shot and killed the first known grizzly-polar bear hybrid to have occurred naturally (a few hybrids have been born among animals in zoos). The creature, dubbed a grolar bear or pizzly, was the result of a mating between a male grizzly and a female polar bear. It is not unusual for a grizzly to wander into polar bear territory—brown bears sometimes stroll onto the pack ice to finish off the remains of polar bear kills. But they don't usually mingle with the locals. The two species are generally aggressive towards each other when they meet, but this odd couple seems to have put aside their differences long enough to mate. The bear looked more like a grizzly than a polar bear; although its coat was white, it was splashed with patches of brown, and the bear had the grizzly's long claws, shoulder hump and dish-shaped face. Scientists suggest that, had it not been shot, the grolar bear could have successfully mated with either bear species.

A Species for Every Niche

The *Lonicera* fly is a hybrid fruit fly found throughout North America. It is thought to be the result of a pairing between the blueberry maggot and the snowberry maggot. Okay, you ask, where is the weird bit? Well, this fruit fly is one of the only known examples in nature where a pairing of two different species created an entirely new species. Also, in its larval stage, the fly feeds on honeysuckle plants that were introduced to our country as ornamental plants, which means that the new species carved out a niche for itself in plants that also did not originally belong in its habitat.

Where's the Beef?

As its name suggests, the beefalo is a hybrid of domestic cattle and bison. Obviously this cross didn't happen without a little help from humans; a rogue bison didn't break into a cattle pen and have his way with the cows, spawning a new breed (there were some instances in the early days of ranching when the two species hooked up on their own, but the resulting offspring were not fertile). Nope, humans and their never-ending quest for the almighty dollar are to blame for this hybrid. Farmers have worked hard to create a fertile hybrid that combined the heartiness of the bison with the

even temper of cattle. The end result is said to be able to withstand harsh North American winters, require little help with calving and be less destructive to grazing lands.

Despite the fact that the meat is reportedly lower in fat and cholesterol than regular beef, you probably won't be chowing down on a beefalo burger anytime soon. The meat has received a lukewarm reception, at best, in the market.

NICE KNOWIN' YA

Nice Canines, Smiley

As soon as I hear the word "Pleistocene," an image of the sabre-toothed cat, or smilodon, taking down a mastodon pops into my head. This large prehistoric cat was a bit smaller than the modern tiger, but it was much stockier in build and could weigh anywhere from 55 to 300 kilograms, depending on the species (of which there were many). The most remarkable feature of this cat, obviously, was the large canine teeth that grew from its upper jaw and could reach a length of 17 centimetres.

The sabre-toothed cat was a large-prey specialist, singling out such animals as bison, American camels, horses and mastodons. However, unlike modern big cats that take down large prey, the smilodon did not kill its victims with a well-placed, suffocating bite to the neck. It couldn't—the large canines would get in the way. Also, this scary-looking predator actually had a relatively weak bite, with only one-third of the power of a lion. In fact, there seems to have been a correlation between jaw strength and canine size: the larger the canines, the weaker the bite. To kill its prey, the big cat most likely overpowered the animal, bringing it to the ground, then bit into its throat, severing the jugular or windpipe with its sabre-like teeth. Its killing style, however, might have been this cat's undoing. Much of North America's megafauna died off as the Pleistocene was drawing to a close. Although they were highly efficient for killing large animals, the prehistoric cat's canines were not suited to catching smaller prey, and its stocky build made it slower than many of the smaller species that survived the mass extinction. By about 10,000 years ago, it too had gone the way of the mammoths.

> **megafauna:** very large animals; usually used to describe undomesticated animals that are larger than a human; most commonly used to refer to mammals of the Pleistocene, but also to mammals such as moose and bears

So Much for Bergmann's Rule

When one thinks of a sloth, one generally pictures the creature dangling from a branch in the rainforests of South and Central America, which the modern-day sloth calls home, not walking about in the Great White North. But at one time, a number of ground sloth species roamed the landscape of what is now Canada. Bergmann's rule states that animals living in cold, northern climates are bigger than their southern counterparts that live in milder climates. Not so with the giant ground sloth (*Megalonyx jeffersonii*), which lived in North America until the end of the Pleistocene, 10,000 years ago. Populations in the northern part of this creature's range were markedly smaller than those that lived farther south. Scientists suggest that this may have been the result of a poor diet, with appropriate vegetation becoming scarcer in the more northerly latitudes, but it is still surprising because one would expect that an animal as slow-moving as a sloth would require a larger body to stay warm in such a harsh climate. Unfortunately, no pelage of the giant sloth has ever been found, so there is no way of knowing how much insulation its coat would have offered. In Canada, fossilized remains of this sloth have been discovered in the Northwest Territories, the Yukon, British Columbia, Alberta and Saskatchewan.

One Nasty Arachnid

Scorpions tend to get a bad rap. They are pretty scary-looking, what with a curled stinger at the back and lobster-like pincers on the front end. But even the deadliest-looking modern scorpion looks positively harmless next to one of its ancestors. The giant sea scorpion could grow to be more than 2.5 metres long, and it wasn't just intimidating to look at—it was also a formidable predator. This fierce arthropod had no natural predators, except other giant sea scorpions. With its naturally scrappy disposition, fights with other sea scorpions were commonplace, and the victor in the battle had no qualms about devouring its competitor. The giant sea scorpion lived during the Devonian period, about 400 million years ago.

DID YOU KNOW?

Alligators in Canada? Yup. Well, more like an alligatorid in what would one day become Canada. *Leidyosuchus canadensis* was a prehistoric, alligator-like creature that lived during the Late Cretaceous period, 100 to 60 million years ago, in what is now Alberta. From the fossil record, we know that it was a medium-sized reptile with a skull measuring about 40 centimetres long.

Taking to the Air

Pterosaurs, the first known aerial vertebrates, dominated the skies between 230 and 65 million years ago. They had a large distribution, encompassing modern-day North and South America as well as Europe. These flying reptiles had massive wings but no feathers, and there has been much debate about how well they could fly. "Pretty darned well" seems to be the correct answer because their remains have been found hundreds of kilometres away from what would have been the nearest land during their time on earth.

Living the Dry Life

Hylonomus lyelli has earned two titles in the history of life on earth: the first reptile and the first animal fully adapted to life on land. And we've got 'em here in Canada—at least in fossilized form. This prehistoric, lizard-like reptile inhabited riverbanks and swampy areas in what is now eastern Canada during the Late Carboniferous period (about 315 million years ago). It grew to about 20 centimetres long and was most likely an insectivore. Fossilized bones of this creature were found in petrified tree stumps in Joggins, Nova Scotia. The United Nations declared the area a World Heritage site in 2008, and Nova Scotia has named *Hylonomus lyelli* its provincial fossil.

AT THE HAND OF HUMANS

Missing Mink

Ever heard of a sea mink? Don't worry—you're not alone. Most people are not familiar with this creature because the species was wiped out before scientists had a chance to study its habits or even classify it taxonomically. Consequently, most of this critter's lifestyle is shrouded in mystery. This is what we do know: the sea mink was a good-sized member of the weasel family, similar in appearance to the American mink but larger and stockier in build. It had coarse, reddish-brown fur and a rather bushy tail. It also, apparently, had a distinctive, rather unpleasant smell. This mustelid was found on rocky coastlines and in the coastal waters of Nova Scotia, New Brunswick and PEI, and it appears to have spent most of its time in the water, where it is thought to have preyed on the Labrador duck, another Canadian species driven to extinction as a result of human carelessness and greed. The sea mink's large size is what ultimately led to its destruction. Trappers in the fur trade could get a pelt that was almost twice the size of an American mink pelt. The largest recorded animal, trapped in Nova Scotia, measured an impressive 82.6 centimetres long (a mink is usually no more than 60 centimetres in length, and a good third of that length is the tail). The last validated sighting of a sea mink was in 1860.

> **mustelid:** a member of the weasel family, Mustelidae, including weasels, mink and ferrets

An Auk-ward Situation

When they saw us coming, the great auks should have beaten a hasty retreat, like any sensible bird would. Instead, these gentle, inquisitive birds hung around to check things out, and they were wiped out for their curiosity. The great auk was a northern Atlantic seabird, found off the coasts of Nova Scotia, New Brunswick and Newfoundland as well as Iceland, Greenland and Britain. With its black back and white belly, it looked rather like a penguin, and it moved about on land in a similar manner. This flightless bird reached 85 centimetres tall and spent most of its time in shallow water, only coming ashore to breed, when it would lay just one egg, not nearly enough to replenish the population as quickly as it was being culled. The bird's numbers were already in decline by the late 1600s because of human overexploitation for its meat, feathers and fat, as well as for specimens for museums and private collections. The last breeding pair was killed in Iceland on June 3, 1844, by collectors.

DID YOU KNOW?

Although it was flightless, the great auk migrated, spending its winters in southern Florida and southern Spain.

ON THE BRINK

Bye Bye Blues

It is difficult to imagine that such an immense creature living in such a remote habitat could be so drastically affected by the behaviour of humans, but the numbers speak for themselves. In 1900, the blue whale population was estimated at 196,000; today there are fewer than 10,000 animals remaining, and their future looks pretty bleak. COSEWIC has this whale listed as endangered.

To put into perspective just how large these whales are, consider this: a Volkswagen beetle could comfortably park in the whale's heart, and the car's driver could exit his vehicle and crawl through the whale's arteries, if he were so inclined.

Shark Attack

It seems odd to me that one of the most successful predators in terms of prey capture should be flirting with extinction, but it's true. Even being at the top of the marine food chain isn't enough to save the great white shark—this huge fish is fighting a losing battle for survival, primarily because of the fin trade. Great white fins are much in demand in many Asian countries and can fetch $700 per kilogram in the market.

Because of its aggressive nature, this shark is an easy target. Just throw a little bait into the water, and the bold predator will swim right up to your boat to snatch it. Also, the great white is an extremely curious creature—individuals have been seen inquisitively nudging boats off the coast of Vancouver.

I guess it's true what they say about your greatest strength also being your greatest weakness—the same boldness and aggression that allows the shark to catch the majority of its prey can also mean that the shark finds itself on the wrong end of a poacher's finning blade.

ABOUT THE AUTHOR

Wendy Pirk

Wendy Pirk has been an animal lover for as long as she can remember. As a child, she dreamed of being a veterinarian but decided instead to pursue a degree in ecological anthropology. She is a serious environmentalist and has worked for the Western Canadian Wilderness Committee and headed the "Stop the Grizzly Hunt Campaign." She now works as a freelance writer and editor with a focus on nature and the environment. This is her second nonfiction book.

ABOUT THE ILLUSTRATORS

Peter Tyler

Peter is a recent graduate of the Vancouver Film School's Visual Art and Design and Classical animation programs. Though his ultimate passion lies in filmmaking, he is also intent on developing his draftsmanship and story-telling, with the aim of using those skills in future filmic misadventures.

ABOUT THE ILLUSTRATORS

Roger Garcia

Roger Garcia is a self-taught artist with some formal training who specializes in cartooning and illustration. He is an immigrant from El Salvador, and during the last few years, his work has been primarily cartoons and editorial illustrations in pen and ink. Recently he has started painting once more. Focusing on simplifying the human form, he uses a bright, minimal palette and as few elements as possible. His work can be seen in newspapers, magazines, promo material and on www.rogergarcia.ca.